Sustainable Geopolymers

This book provides a comprehensive exploration of geopolymers, covering bibliometric analysis, preparation methods, property testing, design modelling, practical applications, and lifecycle assessment. It presents advanced technical frameworks and guidelines for engineers, researchers, academics, and policymakers, promoting sustainable practices aligned with global initiatives such as the Paris Agreement, the Sustainable Development Goals (SDGs), and the African Union Agenda 2063.

The chapters systematically address foundational knowledge, preparation techniques, testing methodologies, design modelling, real-world applications, and lifecycle impacts of geopolymers.

Features

- Provides comprehensive guidelines and strategies for sustainable geopolymer design and practice.
- Explores the integration of artificial intelligence with sustainability frameworks.
- Addresses geopolymer lifecycle assessment in accordance with ISO 14040 and ISO 14044 standards.
- Includes sample tables (e.g., mix design parameters and LCA results) to demonstrate formatting and presentation.
- Focuses on sustainable geopolymers aligned with the SDGs and circular economy principles, supported by case studies.

This book equips readers with the knowledge and tools needed to innovate in geopolymer chemistry and sustainable design.

Sustainable Geopolymers

Properties, Design, and Practice

Jabulani Jolly Matsimbe, Megersa Olumana Dinka,
and David Olatunde Olukanni

CRC Press
Taylor & Francis Group
Boca Raton London New York

CRC Press is an imprint of the
Taylor & Francis Group, an **informa** business

Designed cover image: Getty Images

First edition published 2027
by CRC Press
2385 NW Executive Center Drive, Suite 320, Boca Raton FL 33431

and by CRC Press
4 Park Square, Milton Park, Abingdon, Oxon, OX14 4RN

CRC Press is an imprint of Taylor & Francis Group, LLC

ISBN: 9781041308997 (hbk)
ISBN: 9781041333524 (pbk)
ISBN: 9781003787389 (ebk)

DOI: 10.1201/9781003787389

Typeset in Times
by Newgen Publishing UK

Contents

Authors' Biographies

Jabulani Jolly Matsimbe holds a PhD in Civil Engineering (University of Johannesburg, South Africa), an MSc in Mining Engineering (University of Exeter, UK), and a BSc in Civil Engineering (University of Malawi). He has over ten years of experience as an academic, researcher, and engineer. He teaches geomaterials and geotechnical engineering at the Malawi University of Business and Applied Sciences (MUBAS), where he leads the Sustainable Geomaterials and Geoengineering (SG^3) Research Group and coordinates the Computer Applications for Engineers course. He has also served as coordinator for engineering diploma programmes and student industrial attachments at MUBAS. His research interests span mining waste valorisation, geopolymers and low-carbon cementitious materials, multi-scale geomaterials characterisation, soil stabilisation and ground improvement, geotechnical design, durability and performance-based materials engineering, machine learning and statistical modelling, computational and AI-driven geotechnics, remote sensing, digital geotechnics, circular economy, and sustainable mining systems. He has published over 16 peer-reviewed journal articles, ten international conference papers, and one book chapter; contributed to the ESKOM-University of Johannesburg ash beneficiation and dam rehabilitation research projects; contributed to SAICE technical reports and ethics handbook; and serves as a peer reviewer for several international journals. His academic excellence has been recognised through several prestigious awards, including the Best Paper Award at the ASOCSA 17th Built Environment Conference, the EU-ASIM Scholarship, the Dr. David Livingstone Academic Award (University of Strathclyde–UNIMA) and the World Bank Skills Development Scholarship.

Megersa Olumana Dinka graduated with a PhD from the University of Natural Resources and Applied Life Science (Vienna) in 2010. He also did postdoctoral research at Tshwane University of Technology (2012–2014). He has an MSc in Irrigation Engineering from ArbaMinch University (Ethiopia). He has more than 20 years of experience as an academician and about 24 years of experience as a researcher. Currently, he is Full Professor at the Department of Civil Engineering Science, University of Johannesburg. He has successfully taught various courses and modules at undergraduate and postgraduate levels. Currently, he is teaching Hydrology, Hydraulics and Water Treatment Technology Modules at the University of Johannesburg. Moreover, he has successfully supervised several postgraduate students (35 MSc and 10 PhD). He has published more than 80 journal articles, two books, 15 book chapters, and 25 conference proceedings in accredited publications.

David Olatunde Olukanni is Professor of Civil Engineering with a special interest in Environmental Engineering. His research focuses on Waste Management, Material and Resource Recovery and its Sustainable Application, Lifecycle

Assessment, Environmental Sustainability, Circular Economy, Waste to Energy, and Wastewater Treatment Systems. He is the pioneer Chair of the Waste to Wealth Initiative and Centre for Research in Waste Management at Covenant University. He is a Fulbright Fellow to the United States and has won several awards and grants for research and travel scholarships. He is a member of learned and professional associations: COREN, NSE, ASCE, IAEH, ICCWtE, Chair, country-specific working group of IconSWM-CE. He is a member of the Council of COREN and chairs the committee that oversees two departments: Planning, Research & Statistics and Quality Assurance & Compliance. He served as Head of the Civil Engineering Department; Director, Quality Assurance and Academic Standards; Pioneer Director, Covenant University Central Instrumentation Research Facility (CUCIRF); Dean, College of Engineering; and Chair, Senate Business Committee of Covenant University. He is a Visiting Scholar and Researcher to the Department of Civil, Construction and Environmental Engineering at NCSU, USA and MUBAS, Malawi. He has authored over 180 peer-reviewed articles, proceedings and invited papers and has done several consultancy projects. He has supervised many Masters' and Doctoral Degree students. He serves as assessor and external examiner for postgraduate dissertations and theses in many universities. He is the cluster head of Urban Environmental Issues and National Development at Covenant. He has international collaborations in many countries: USA, UK, Canada, China, Germany, Spain, France, India, South Africa, Egypt, Tanzania, Malawi, Zambia, and Congo, to name a few. He is an Ordained Pastor in the Living Faith Church.

Preface

Civil engineering structures are typically designed and constructed using ordinary Portland cement, steel, aluminium, soil, rock, and wood, adhering to codes and standards that have been extensively researched and understood for over 200 years. These standards have been fully developed for practical application by the construction industry. However, sustainable geopolymer design and practice represent a novel field that lacks established standards and practices, as it has only been studied for less than 50 years. To address this gap and to build industrial confidence, as well as broad acceptance among engineers and construction practitioners, a comprehensive book is needed that provides guidelines and strategies for sustainable geopolymer design and practice, tailored to meet complex, multifunctional, and specialized engineering applications. This book aims to explore the rapidly evolving field of geopolymers, covering a wide range of topics from theoretical insights to design and professional practices aligned with the circular economy and the UN Sustainable Development Goals (SDGs) 9, 11, 12, 13, and 17. The integration of artificial intelligence (AI) and machine learning (ML) approaches in geopolymer design is particularly compelling.

The book covers geopolymer bibliometric analysis, geopolymer preparation, geopolymer properties testing, geopolymer design modelling, geopolymer practical applications, and geopolymer lifecycle assessment. This book enhances industry competitiveness through sustainable innovative practices, waste valorisation, digital innovation, skills development, and standardisation. Furthermore, it provides an advanced technical reference and roadmap for practicing engineers, engineering students, researchers, policymakers, digital innovation entrepreneurs, and construction practitioners to make informed decisions towards capacity building and the broader adoption of geopolymers and AI, promoting the digital transformation, circular economy, UN SDGs 9, 11, 12, 13, and 17, the International Energy Agency's GlobalABC Roadmap for Buildings and Construction of 2020–2050, the UN Climate Change Conference of the Parties Paris Agreement, the Intergovernmental Panel on Climate Change, the African Union Agenda 2063, the Malawi Vision 2063, and the South African National Development Plan 2030. The fundamental principles of sustainable geopolymer design and practice can specifically assist practicing engineers, engineering students, construction practitioners, and researchers in becoming better civil engineering material designers, helping those seeking to enhance their existing knowledge of Geopolymer Chemistry and emerging as leaders in Sustainable Geopolymer Design and Practice, coupled with AI and machine learning.

Chapter 1 aims to provide readers with a systematic understanding of the knowledge base of geopolymers. Readers will be able to analyse complex issues systematically and make sound judgements in civil engineering situations requiring knowledge of geopolymer materials.

Chapter 2 aims to provide readers with knowledge on preparing geopolymers for civil engineering projects. Readers will be able to choose the appropriate precursor materials and develop geopolymer paste, mortar, or concrete for structural applications.

Chapter 3 aims to provide readers with scientific knowledge of the laboratory-based testing methods for geopolymers. Readers will be able to select the appropriate testing technique for assessing specific geopolymer properties necessitating acceptable engineering performance and practice.

Chapter 4 aims to provide readers with mathematical knowledge underpinning the engineering principles and tools required for developing geopolymer mix design models. Readers will be able to apply various scientific modelling approaches and AI/ML to the mix design of geopolymers and innovatively justify their selection based on acceptable performance.

Chapter 5 aims to inform readers of the real-world impact of adopting geopolymers on engineering practice. Readers will be able to demonstrate a deeper understanding of the relationship between geopolymer design and practice, supported with case studies.

Chapter 6 aims to provide readers with knowledge of the lifecycle of geopolymer development and applications. Readers will be able to demonstrate a deeper understanding of carbon emissions and identify environmental impacts and ethical responsibilities in accordance with acceptable geopolymer practice.

Jabulani Jolly Matsimbe
Megersa Olumana Dinka
David Olatunde Olukanni

1 Geopolymer Bibliometric Analysis

1.1 EVOLUTION OF GEOPOLYMERS

In this era of the circular economy, waste management, sustainable cement, zero-cement construction systems, and the achievement of Sustainable Development Goals (SDGs) 9, 11, 12, and 13, engineers and construction practitioners must adopt geopolymers as an alternative and/or supplement to ordinary Portland cement (OPC). The continued growth in the use of OPC (i.e., over 4 billion tonnes/year) severely impacts the environment through CO_2 emissions, which currently accounts for 8% of global greenhouse gas emissions (Matsimbe et al., 2023). Approximately 2 tonnes of raw materials are consumed to produce 1 tonne of OPC, resulting in the emission of 1 tonne of CO_2 during the process. The growth in the use of OPC has been attributed to extreme urbanization, population growth, and infrastructure development. Geopolymers, an invention of Professor Joseph Davidovits, offer comparable physico-chemo-mechanical performance with a reduction in CO_2 emissions by 45% compared to OPC (Matsimbe et al., 2022a). Geopolymer production is a green technology whereby a natural or waste aluminosilicate precursor material chemically reacts with an alkaline hydroxide and/or alkaline silicate to produce a geopolymer paste, mortar, or concrete (Provis and van Deventer, 2009). When the aluminosilicate precursor is exposed to an alkaline solution, it undergoes dissolution, nucleation, oligomerization, polymerization, and hardening to form the geopolymer network (Davidovits, 2020). The commonly used natural precursor consists of metakaolin, laterite, or volcanic ash, while the waste precursor consists of fly ash, ground granulated blast furnace slag, phosphogypsum, rice husk ash, silica fume, red mud, mine tailings, or palm oil fuel ash. Due to the promotion of SDG 13, sustainable waste management, and circular economy, waste precursor materials are commonly championed to produce geopolymers for construction in line with the UN Climate Change Conference of the Parties, the Intergovernmental Panel on Climate Change, and the International Energy Agency GlobalABC Roadmap for Buildings and Construction (IEA, 2020).

Overall, the geopolymer field has matured from foundational work, such as chemistry and activation theory, to application-oriented projects, including geopolymer concrete, composite, and waste stabilisation. The emergence of

DOI: 10.1201/9781003787389-1

durability, sustainability, lifecycle assessment (LCA), and 3D printing shows an increasing concern for real-world performance and environmental claims. Geopolymers have successfully been used to construct the University of Queensland Global Change Institute Building, Brisbane West Wellcamp Airport, Melton Library/Learning Hub, West Gate Freeway upgrade, Salmon Street Bridge, and South African City Deep container terminal (Matsimbe et al., 2022b). Embracing geopolymer science and technology in the Fourth Industrial Revolution (4IR) enhances industry competitiveness through sustainable innovative practices, proactive waste materials management, portfolio diversification, and skills development, ultimately promoting global socio-economic welfare and societal impact.

1.2 RESEARCH TRENDS IN GEOPOLYMERS

Compared to systematic literature reviews (SLR), the bibliometric analysis technique creates quantitative visualisation network maps that highlight growth patterns and projections in geopolymer research. Bibliometric analysis uses tools such as VOSviewer, CiteSpace, and Bibliometrix to map intellectual structures, identify research hot spots, and visualise scientific networks. Table 1.1 provides

TABLE 1.1
Differences between bibliometrics and systematic literature review

Aspect	Bibliometric Analysis	Systematic Literature Review
Purpose	Quantify, map, and visualise patterns in a research field	Synthesise and critically evaluate existing research findings
Approach	Quantitative through statistics, networks, and trends	Qualitative and interpretive through thematic synthesis and critical appraisal
Questions answered	How much? How often? Who collaborates? What trends exist?	What does the evidence say? What are the gaps? How strong is the evidence?
Data sources	Large datasets from Scopus, Web of Science, and Google Scholar	A curated set of studies selected through strict inclusion and exclusion criteria
Tools	VOSviewer, CiteSpace, Bibliometrix	PRISMA framework, thematic coding, and meta-analysis tools
Output	Maps, graphs, clusters, networks, citation metrics	Conceptual frameworks, thematic summaries, evidence-based conclusions
Nature of analysis	Macro-level, i.e., big picture trends	Micro-level, i.e., deep understanding of specific studies
Timeframe	Covers entire bodies of literature over decades	Focuses on fewer, higher-quality studies

a detailed description of the differences between bibliometrics and systematic literature reviews. Literally, bibliometrics can be seen as an aerial view of the entire forest (i.e., patterns, clusters, density) while systematic literature review can be symbolised as walking through the forest and examining each tree closely.

Bibliometric analyses aim to uncover how knowledge is produced, organised, and evolves within a research domain. Two of the most powerful techniques for achieving this are co-citation analysis and keyword co-occurrence analysis. Combined, they help map both the intellectual foundation and the research front of a field. Co-citation occurs when a third document cites two documents together. For example, if documents A and B frequently appear together in citation lists, they are considered intellectually related. Co-citation reveals:

- *Intellectual structure of the field:* Co-citation analysis identifies core theories, foundational papers, and major schools of thought. It helps determine which authors or papers form the intellectual backbone of the field, what theoretical traditions dominate, and how influential works have shaped subsequent research.
- *Clusters of foundational knowledge:* Highly co-cited documents group into clusters representing research traditions, landmark contributions, and methodological paradigms. For example, a geopolymer study may have clusters in cement chemistry, material durability, or alkali activation literature.
- *Historical development:* By tracking co-citation over time, one can trace how concepts emerge, how theories diverge or converge, and shifts in dominant paradigms.

In contrast, keyword co-occurrence analysis examines how frequently two or more keywords appear together within the same article. Keywords reflect the content of publications; co-occurrence shows thematic connections. Keyword co-occurrence reveals:

- *Thematic structure of the field:* This approach identifies dominant themes, emerging topics, and interdisciplinary linkages.
- *Research front:* While co-citation maps the intellectual foundation, keyword co-occurrence maps the current trends, i.e., active areas where new knowledge is being generated.
- *Hotspots and emerging trends:* High-frequency or rapidly increasing keyword clusters indicate new topics gaining attention, areas where research investment is rising, and future directions of the field (e.g., low-carbon cement, geopolymer concrete, alkali activation, and machine learning).
- *Conceptual relationships:* Co-occurrence clusters show how concepts are related in practice, e.g., geopolymer concrete co-occurring with durability and fly ash may form one cluster. Similarly, circular economy co-occurring with sustainable materials forms another.

From the foregoing, it can be observed that keyword co-occurrence is crucial to the identification of research trends and their interdisciplinary nature. Figure 1.1 demonstrates the keyword mapping for "geopolymer" in the Scopus database. This is accomplished by exporting CSV data files from the Scopus database and analysing or visualising them using VOSviewer software developed by van Eck and Waltman (van Eck and Waltman, 2010). A typical search string targets title/abstract/keywords for terms like "geopolymers" *or* "inorganic polymer" and restricts by document type (articles/reviews), language (usually English), and date range (e.g., 2020–2026). The trending domains for advancing knowledge in geopolymers are illustrated through keyword co-occurrence. The visualisation consists of four clusters: cluster 1 (geopolymer, compressive strength, fly ash, concrete mixtures, concrete aggregates, Fourier transform infrared spectroscopy, and demolition); cluster 2 (geopolymer concrete, geopolymer composites, fracture mechanics, sustainable development, sustainability, binders, and fibre-reinforced concrete); cluster 3 (bending strength, bending moment, tensile strength, mortar, silica fume, silicates, X-ray diffractions, mechanical, and sodium silicate); and cluster 4 (Portland cement, waste glass, glass fibres, slag, ground granulated blast furnace slag, and soil stabilisation). Geopolymer is a topical keyword indicating the broad classification of the research field. The remaining keywords identify specific knowledge components that facilitate search and retrieval functions.

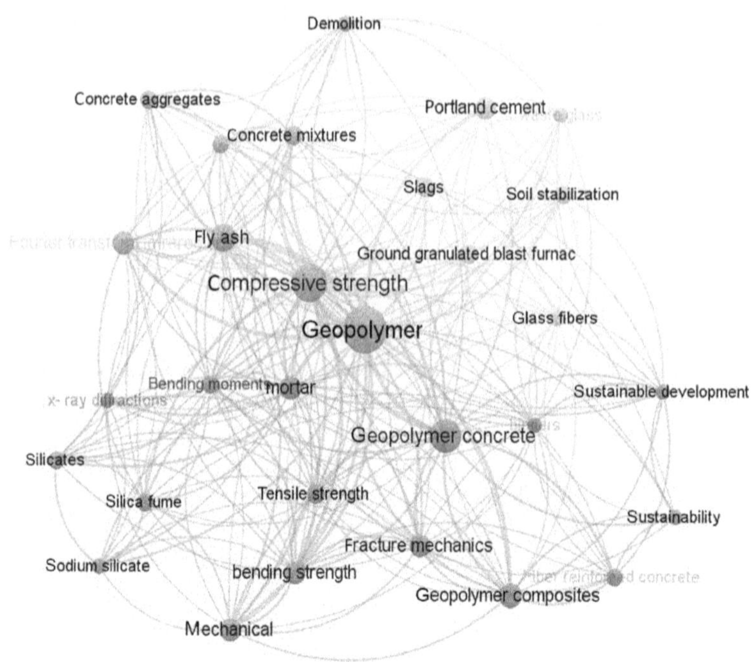

FIGURE 1.1 Keywords co-occurrence visualisation.

Visual analysis of the keywords helps engineers, construction practitioners, and researchers predict the thematic directions for enhancing research and practice in the geopolymers domain.

Depending on the aggregate type (i.e., fine or coarse), the binder produced can be a geopolymer paste, mortar, or concrete. Nevertheless, the most produced material is geopolymer concrete due to the need for heavy structural support in construction. The geopolymer paste and mortar can also be utilised for making bricks, sealing crack joints, and repairing concrete. The fracture mechanics of geopolymers is crucial to understanding the stress and strain of geopolymer products when exposed to an external load (Matsimbe et al., 2022b). This understanding adds weight to the development of geopolymer standards for engineering applications. Despite the numerous precursor materials, the bibliometric network shows that fly ash, slags, ground granulated blast furnace slag, demolition waste, and waste glass are considered the go-to precursor waste materials. This trend can be attributed to the increased demand for energy, steel, and glass for modern infrastructural development. The precursor waste materials are promoted due to the growing need for environmental protection, a circular economy, and adherence to SDG 13. OPC, with an 8% CO_2 emission (Matsimbe et al., 2023), is often considered a reference benchmark material against which other alternative materials are compared and proposed as supplementary cementitious materials.

Understanding the physico-chemo-mechanical properties of a geopolymer enhances industry confidence in its selection as a construction material (Matsimbe et al., 2024b). The commonly studied mechanical properties consist of compressive strength, tensile strength, bending strength, bending moment, and fracture mechanics. To enhance the tensile strength of geopolymers, fibre reinforcement is introduced into the system, resulting in fibre-reinforced geopolymer concrete. The microstructure of a geopolymer is crucial to its mechanical performance. This internal arrangement influences the material's strength, hardness, and durability (Matsimbe et al., 2024a). For example, controlling the grain size can change a geopolymer's strength and robustness. The commonly studied microstructural properties encompass mineralogy, morphology, chemical composition, and bonding, which are evaluated using X-ray diffraction (XRD), scanning electron microscopy energy dispersive spectroscopy (SEM-EDS), X-ray fluorescence (XRF), and Fourier transform infrared (FTIR), respectively. All these geopolymer concepts and applications can be graphically visualised to predict their growth trends and outlook. Figure 1.2 illustrates that the regression of the geopolymer research trend is significant, as indicated by an R^2 value of 0.9154.

The polynomial equation for the research trendline can be used to predict the probability of growth in geopolymer research. The concepts of sustainability, sustainable development, and circular economy underpin the entire range of geopolymer development (Matsimbe et al., 2023). This is because geopolymer is a sustainable green technology that utilises waste materials and an alkaline solution for its creation, reducing reliance on OPC and natural raw material extraction, which is detrimental to the environment. The increased awareness of global warming and

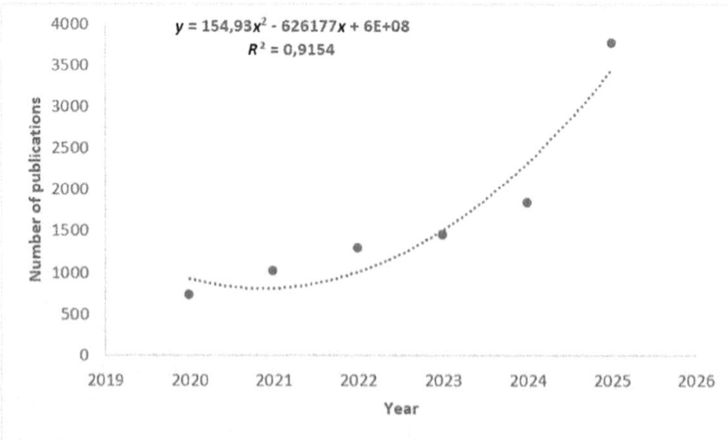

FIGURE 1.2 Statistical regression of the publications trend for geopolymer.

CO_2 emissions issues has urged the construction industry to research greener and/or eco-friendly alternative cements to OPC in line with the circular economy and sustainable development goals (SDGs) 9 (Industry, Innovation, and Infrastructure), 11 (Sustainable Cities and Communities), 12 (Responsible Consumption and Production), and 13 (Climate Action).

Geopolymer research has grown substantially since the 1970s, and bibliometric trends clearly indicate a shift from foundational chemistry studies to sustainability-driven and application-oriented research (Matsimbe et al., 2022a). This transformation mirrors the broader global push for low-carbon construction materials and circular economy practices. The early phase (1970–1990) focused on advancing fundamental geopolymer chemistry and material science foundations. Aluminosilicate reactions, polymerisation mechanisms, the role of alkali activators, and microstructure development were the leading research topics that dominated in the early phase. The middle phase (2000–2010) focused on optimisation, durability, and engineering applications. Geopolymer research shifted from laboratory curiosity to usable construction materials. There was an emergence of clusters in fly ash geopolymer, durability, mechanical properties, and engineering applications. The mature phase (2010–2023) focused on promoting sustainability, the circular economy, and low-carbon technologies. Geopolymers are now viewed as strategic critical materials for decarbonising the cement industry, with strong clusters around lifecycle assessment, carbon footprint, and industrial waste reuse. This book introduces and/or advances the digital innovation phase (2024–present), which integrates all prior phases of geopolymer research into artificial intelligence-driven and machine learning transformation.

The understanding of which countries are most active in each research area helps to strengthen future partnerships, encourage technology transfer, and attract joint research funding. Figure 1.3 illustrates that China, India, the United States,

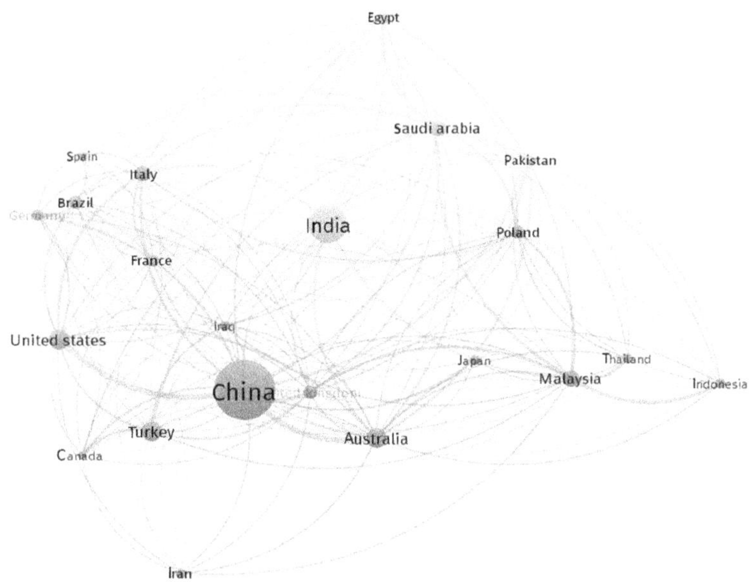

FIGURE 1.3 Visualisation of countries leading in geopolymer research.

and Australia are the leading countries in geopolymer research and development. This suggests that both economically developed nations and developing economies such as India recognise the potential of geopolymers and are actively investigating their use in construction. Collaboration mapping further shows strong research connections among India, Australia, and China.

Within the African continent, countries such as Egypt, Cameroon, Nigeria, Morocco, and South Africa have demonstrated tremendous potential as emerging geopolymer researchers supporting sustainable infrastructure development in Africa (Matsimbe et al., 2022a; Matsimbe et al., 2023). Given the continent's reliance on cement and concrete technologies, Africa remains an attractive destination for major global economies. Africa's attractiveness lies in the convergence of market size, growth potential, raw materials, and opportunities in green technology. The need for housing, roads, bridges, ports, and energy infrastructure translates directly into sustained demand for cement and concrete. By 2050, Africa's urban population is projected to double, driving long-term requirements for construction materials. Global investment initiatives, such as the African Continental Free Trade Area (AfCFTA), promote intra-African trade and logistics infrastructure, further fuelling cement consumption. Africa possesses vast reserves of fly ash, slag, and phosphogypsum from emerging industries serving as critical feedstocks for low-carbon binders like geopolymers. As global economies shift toward carbon-neutral construction and NetZero, Africa provides opportunities to leapfrog directly to greener cementitious technologies. For advanced global economies, investing in cement and concrete technologies in Africa:

- secures long-term market access;
- aligns with global carbon-reduction strategies and sustainable supply chains;
- promotes technology transfer and research collaborations;
- supports the UN Sustainable Development Goals.

The mapping of geopolymer networks reveals collaboration pathways, e.g., national clusters, regional hubs, and international bridges. Engineering practitioners and policymakers can use these geopolymer networks to identify emerging markets for investment, develop capacity-building initiatives, and foster cross-regional collaborations. Bibliometric analyses of geopolymer literature give a robust, data-driven picture of a maturing field that is shifting from laboratory exploration toward application, durability assessment, and sustainability evaluation.

1.3 ADVANTAGES OF GEOPOLYMER ADOPTION

The research, development, and adoption of geopolymers offer numerous society-informed benefits over ordinary Portland cement in the following ways:

- Reduction in energy demand and global warming will be due to minimal release of CO_2 emissions. Geopolymer is no longer viewed as a concept for a greener society but rather as a pragmatic solution for reducing CO_2 emissions in the construction industry. It is commonly manufactured as a geopolymer binder, mortar, brick, and concrete.
- Promoting the reuse of industrial waste materials advances the circular economy and lowers environmental pollution. Geopolymer provides a sustainable low-carbon construction material that promotes circularity through the high-volume valorisation of waste materials. It further protects the environment from waste disposal in landfills and promotes the advancement of SDGs 9, 11, 12, and 13.
- Geopolymers broaden the market revenue for greener, low-carbon cement, thereby advancing sustainable infrastructure development. Protecting the environment from natural resource extraction and waste landfills that pollute soil and water systems is a key sustainability enabler.
- The three-dimensional geopolymer network structure provides superior mechanical strength, fire resistance, thermal insulation, minimal shrinkage, and strong durability in aggressive environments. These superior performance properties make the geopolymer a sustainable cementitious alternative in construction and environmental sustainability engineering, thus finding versatile applications in radioactive waste immobilisation, heavy metal and leachate adsorption, soil stabilisation, ground improvement, and water and wastewater treatment.

Overall, geopolymers excel in environmental sustainability, durability, and chemical and fire resistance. Furthermore, they offer high early strength and

TABLE 1.2
Comparative advantages of geopolymer over OPC

Property	Geopolymer	OPC
CO_2 emissions	Reduced by over 45%	High due to clinker production
Raw materials	Industrial waste: fly ash, slag, rice husk ash, phosphogypsum	Natural: limestone, clay, gypsum; energy-intensive processing
Early strength	High early strength with heat or ambient curing adjustments	Moderate early strength
Durability	Excellent chemical, sulphate, and acid resistance	Moderate; vulnerable to sulphates, acids, and chlorides
Fire resistance	Maintains integrity at high temperatures	Prone to spalling and cracking at high temperatures
Permeability	Low, reduced water and chloride penetration	Higher, more prone to water and chloride ingress
Cost considerations	Potentially cheaper with local waste materials	Usually higher due to clinker production and transportation
Environmental impact	Utilises waste, reduces landfill, and is sustainable	Extractive, high energy, high CO_2 footprint

long-term cost advantages, making them a strong alternative to ordinary Portland cement, especially in aggressive environments and sustainable construction projects (Elchalakani et al., 2023). Table 1.2 provides a detailed summary of the uniqueness of geopolymers compared to OPC. The suggested properties are discussed in detail in subsequent chapters.

Overall, geopolymers have evolved as a greener, eco-friendly third-generation binder alternative to OPC, as they exhibit comparable physico-chemo-mechanical performance, low CO_2 emissions, and reduced embodied energy, making them suitable for sustainable construction. Sustainability is considered in terms of the cost/benefit ratio; therefore, sustainable cement should have maximum performance and durability (benefit) with a minimum environmental impact (cost). Looking back to move forward, the construction industry must shift from the exploitation of natural virgin materials to recycling and/or reusing waste materials. The sooner all key construction stakeholders are involved in geopolymer science and technology, the better it will be for the built environment in its strive to decarbonise.

PROBLEMS

1. Define bibliometrics and explain how it differs from a systematic literature review.
2. Explain the role of co-citation and keyword co-occurrence analyses in mapping the intellectual structure of a field.

3. Describe how publication trends reflect the maturation of geopolymer science from chemistry-focused research to sustainability-oriented studies and digital transformation.
4. You observe that 60% of geopolymer publications originate from Asia, 20% from Europe, 10% from North America, and 10% from Africa and Latin America combined.
   - What conclusions can you draw about global research capacity?
   - Suggest two strategies to strengthen research participation in underrepresented regions.
5. A bibliometric study reports "China has the highest publication volume in geopolymer research."
   - What further analyses, rather than volume, could you perform to determine impact?
   - Discuss how policy or funding contexts might explain these trends.
6. Evaluate how bibliometric findings could guide funding agencies and universities in shaping national research priorities for sustainable construction materials.

REFERENCES

Davidovits, J. (2020). *Geopolymer Chemistry and Applications* (5th ed.). Institut Géopolymère, Saint-Quentin, France.

Elchalakani, M., Yang, B., Mao, K., and Pham, T. (2023). *Geopolymer Concrete Structures with Steel and FRP Reinforcements*. Woodhead Publishing Ltd, Cambridge, UK.

IEA. (2020). *GlobalABC Roadmap for Building and Construction 2020–2050: Towards a Zero-Emission, Efficient and Resilient Buildings and Construction Sector*. International Energy Agency Technology Report.

Matsimbe, J., Dinka, M., Olukanni, D., and Musonda, I. (2022a). A Bibliometric analysis of research trends in geopolymer. *Materials*, vol. 15, no. 19.

Matsimbe, J., Dinka, M., Olukanni, D., and Musonda, I. (2022b). Geopolymer: A systematic review of methodologies. *Materials*, vol. 15, no. 19.

Matsimbe, J., Dinka, M., Olukanni, D., and Musonda, I. (2023). Bibliometric trends of geopolymer research in Sub-Saharan Africa. *Materials Today Communications*, vol. 35.

Matsimbe, J., Dinka, M., Olukanni, D., and Musonda, I. (2024a). Durability properties of ambient-cured fly ash-phosphogypsum blended geopolymer mortar in terms of water absorption, porosity, and sulfate resistance. *Discover Sustainability*, vol. 5, no. 330.

Matsimbe, J., Dinka, M., Olukanni, D., and Musonda, I. (2024b). Performance evaluation and mix design of ambient-cured fly ash-phosphogypsum blended geopolymer paste and mortar. *Results in Engineering*, vol. 24, no. 103280.

Provis, J., and van Deventer, J. (2009). *Geopolymers: Structures, Processing, Properties and Industrial Applications*. Woodhead Publishing Ltd, Cambridge, UK.

van Eck, N. J., and Waltman, L. (2010). Software survey: VOSviewer, a computer program for bibliometric mapping. *Scientometrics*, vol. 84, no. 2, 523–538.

2 Geopolymer Preparation

2.1 PRECURSOR MATERIALS

To prepare a geopolymer, there is need for a raw material that must react with the activator. The raw material can be either naturally occurring (i.e., metakaolin, laterite, volcanic ash, and zeolite) or an industrial waste (i.e., fly ash, slag, phosphogypsum, etc.). The use of industrial waste materials over natural virgin materials as geopolymer precursors is advocated as it is deemed environmentally sustainable. This is attributed to the depletion of natural resources, increased awareness of CO_2 emissions, and the increase in circularity in solid waste management. The properties of geopolymer concrete or mortar depend strongly on the quality and nature of the precursor materials (Duxson et al., 2007; Provis and van Deventer, 2009). Proper selection of source materials and activators is crucial. Aluminosilicate precursors are the backbone of any geopolymer formulation.

The most used aluminosilicate precursor is fly ash (FA), a by-product of coal combustion captured by electrostatic precipitators in thermal power plants, due to its high amorphous silica and alumina content. At most, 1 billion tonnes of FA are produced annually, and more than 60% is disposed of in landfills, posing environmental hazards. Due to the differences in coal types, coal blending, burning temperature, and boiler performance, FA varies in its chemical composition and reactive glassy phases. This, in turn, leads to varying dissolution mechanisms, polymerisation processes, and hardened engineering properties (Matsimbe et al., 2024; Hardjito and Rangan, 2005). The top contributors to FA production worldwide are China, India, United States, Australia, and South Africa, primarily due to their abundant coal seams and industrialization (Matsimbe et al., 2023). The ASTM C618 standard (ASTM C618, 2022) classifies FA into low-calcium class F and high-calcium class C. The low-calcium class F FA (from anthracite coal) contains a SiO_2+Al_2O_3+Fe_2O_3 content of 50% and CaO less than 18%, while the high-calcium class C FA (from lignite coal) contains a SiO_2+Al_2O_3+Fe_2O_3 content of 50% and CaO more than 18%. The contents and ratios of SiO_2, Al_2O_3, Fe_2O_3, and CaO determine the reactivity of FA and possible performance as a precursor material.

DOI: 10.1201/9781003787389-2

The morphology of FA (see Figure 2.1) is typically spherical, with small, fine particulates, which gives it an advantage over other precursor materials. The spherical ash particles enhance paste workability and reduce water demand, thereby increasing the volume of calcium silicate hydrate (CSH) in the binder and improving its strength. The morphology of fly ash is governed by its spherical particle shape, glassy surface, and fine size distribution, which play a decisive role in its suitability for geopolymer production. Fly ash with high amorphous content, spherical particles, and fine particle size (<75 μm) is ideal for high-performance geopolymer concrete (Matsimbe et al., 2025). The spherical morphology provides a "ball-bearing" effect, which enhances workability in both geopolymer and cement pastes by reducing internal friction between particles. Understanding and characterising these morphological features enable engineers and researchers to control the rheology, setting behaviour, and ultimate strength of geopolymer systems, contributing to the development of more durable and sustainable construction materials.

Bottom ash (BA) is another crucial by-product of coal combustion processes from thermal power plants. As global reliance on coal-fired power remains significant worldwide, managing the resultant ash materials has become an important environmental and industrial concern. BA is the non-combustible residue collected at the bottom of boilers after coal combustion. Unlike fly ash, which remains suspended in flue gases and is captured using air pollution control systems, BA is heavier and remains in the furnace bottom and/or is discharged via a water-based removal system (Matsimbe et al., 2022). Bottom ash is largely

FIGURE 2.1 Morphology of FA observed under SEM.

composed of SiO_2, Al_2O_3, CaO, and Fe_2O_3. The morphology of bottom ash (see Figure 2.2) contains coarse, granular, and irregularly sized particles. BA particles exhibit a heterogeneous and irregular morphology due to incomplete fusion and variable cooling rates during combustion. In contrast to the smooth, spherical particles typical of fly ash, bottom ash particles are generally angular, porous, and rough-surfaced, often with vesicular (bubble-like) textures caused by trapped gases during solidification.

The key morphological features of BA include:

- irregular angular grains, which are often fractured or jagged in shape;
- porous and vesicular structures, with internal cavities formed by the rapid release of gases;
- glass-like fragments interspersed with crystalline regions;
- dark grey to black coloration, depending on unburned carbon content.

The rough and irregular morphology of BA enhances mechanical interlocking in geopolymeric matrices but typically reduces workability and increases water demand compared to spherical fly ash.

Phosphogypsum (PG) is another commonly used precursor material. It is an industrial waste formed from the reaction between phosphate ore and sulfuric acid during the production of fertilizer. At most, 300 million tonnes of PG are generated annually, with China, India, United States, and South Africa as the leading producers. The U.S. EPA and EURATOM permit the potential uses of phosphogypsum in construction and soil stabilisation within the set safe limit of 370 Bq kg^{-1} or 10 pCi g^{-1} (Matsimbe et al., 2024). The chemical composition of PG comprises over

FIGURE 2.2 Morphology of BA observed under SEM.

30% CaO and less than 4% SiO_2, Al_2O_3, and Fe_2O_3, making it a suitable calcium-source precursor material for geopolymer production. Figure 2.3 shows that the morphology of PG particles is typically irregularly shaped crystals with a tendency to form plate-like, needle-like, or prismatic structures. The particle morphology is primarily determined by the crystallisation environment (i.e., temperature, pH, and impurities in the acidulation process), cooling rate of the reaction slurry, and post-processing or washing treatments. Most PG particles exhibit a crystalline structure of calcium sulphate dihydrate ($CaSO_4 \cdot 2H_2O$), commonly referred to as gypsum. These crystals often form aggregates or clusters, leading to high specific surface areas and porous microstructures.

Ground granulated blast furnace slag (GGBFS) is a by-product derived from the iron and steel manufacturing industry. It forms through the rapid quenching and grinding of molten blast furnace slag. GGBFS is commonly used in geopolymer production due to its pozzolanic and latent hydraulic properties. The use of GGBFS in geopolymers not only improves mechanical and durability performance but also contributes to sustainable development through the reduction of CO_2 emissions associated with Portland cement production. GGBFS is rich in CaO, SiO_2, Al_2O_3, and MgO, such that the high glass content ensures the reactivity required for cementitious behaviour (Provis and van Deventer, 2009). GGBFS acts both as a filler and a reactive material whereby it hydrates in the presence of water and an activator, and the SiO_2 and Al_2O_3 in the GGBFS react with $Ca(OH)_2$ to form C-S-H, key to strength development.

Figure 2.4 shows that GGBFS is angular, irregularly shaped particles, which favour mechanical interlocking and enhanced cement-aggregate bond. The morphology of GGBFS particles is primarily governed by the rapid quenching

FIGURE 2.3 Morphology of PG observed under SEM.

FIGURE 2.4 Morphology of GGBFS observed under SEM.

process used to form slag granules and by subsequent grinding conditions. The molten slag, when rapidly cooled by high-pressure water jets (granulation), solidifies into a glassy, granular material with a sand-like texture. After grinding to a fine powder, GGBFS exhibits:

- angular to subangular particle shapes, reflecting fracture surfaces produced during milling;
- irregular and sharp-edged grains under high magnification;
- smooth or slightly rough glassy surfaces, depending on grinding efficiency and glass content;
- dense, solid particles with low internal porosity.

The angular morphology of GGBFS promotes strong interparticle friction, influencing workability in geopolymer and blended cement pastes. The glassy nature of most GGBFS particles provides a reservoir of reactive amorphous aluminosilicate and high-calcium systems (Elchalakani et al., 2023).

Mine tailings (MT) are the finely ground residual slurry materials left over after the extraction of minerals and metals from ore. These by-products emanating from mining operations represent one of the most voluminous and persistent environmental issues faced by the global extractive industry. Tailings management is critical due to the potential for environmental degradation, human health impacts, and catastrophic failures of storage facilities. However, the advancements in geotechnology and circular economy practices have led to the reuse of MT as a secondary source of construction materials, most notably as precursors for geopolymers (Matsimbe et al., 2022). The composition of MT depends on the ore type, beneficiation process, and location; however, the typical constituents include SiO_2, Fe_2O_3, Al_2O_3, CaO, MgO, and SO_3. The tailings are typically stored in dams or ponds and commonly consist of tailings from gold, copper, and iron mines. With innovation and responsible stewardship, the mining industry can transform its legacy tailings into a supply chain for a greener, more circular future.

Rice husk ash (RHA) is a highly pozzolanic material obtained from the controlled combustion of rice husks – a major agricultural waste generated during rice milling. With over 700 million tonnes of rice produced annually worldwide, significant quantities of rice husk are generated, approximately 20% of the paddy weight. With correct burning, rice husks yield ash rich in amorphous silica, suitable for use as precursors in geopolymers. RHA supports sustainable development by reducing landfill waste, lowering greenhouse gas emissions, and improving the performance of cement-based systems. The presence of a high SiO_2 content in the RHA makes it suitable as a pozzolanic additive.

2.2 ALKALINE CATALYST

The type and concentration of the alkaline catalyst play a crucial role in influencing the geopolymerisation process and performance of the hardened geopolymer. NaOH is widely used in geopolymer synthesis due to its strong alkalinity, affordability, and availability. It breaks down the Si-O-Si and Al-O-Si bonds in aluminosilicate precursors, forming soluble silicate and aluminate species that react to form the geopolymer gel (Matsimbe et al., 2024). Notable effects of NaOH include enhancing the early dissolution of precursor materials and accelerating the kinetics of polymerisation.

KOH performs a similar function to NaOH but offers different ionic mobility and dissolution characteristics. K^+ often results in higher strength and denser matrices in some geopolymer systems, particularly those based on slag (Matsimbe et al., 2022). The use of KOH promotes better long-term strength development and lower viscosity than NaOH solutions at equal molarities. Na_2SiO_3 is typically used in combination with NaOH to provide additional silicate species that are essential for the formation of the geopolymeric gel. The SiO_2/Na_2O ratio (silicate modulus) is a critical parameter influencing gel structure and strength. Generally, the use of Na_2SiO_3 enhances gel formation and matrix densification, improves mechanical strength and durability, and affects setting time based on its modulus. Sodium-based activators are more economical, while potassium-based activators yield slightly higher mechanical and thermal stability.

The concentration of hydroxide solutions significantly affects the dissolution of precursors and the rate of gel formation. The use of low molarity (i.e., 4–6 M) leads to incomplete dissolution, a weak gel, and lower strength. Moderate molarities (i.e., 8–12 M) provide optimal dissolution and polymerisation and higher strength. This optimal molarity range usually balances strength and workability. In contrast, the use of high molarity (i.e., >14 M) leads to rapid setting, high viscosity, and microcracking due to excess alkali. In addition, the Na_2SiO_3/NaOH ratio affects the geopolymer matrix's structure (Matsimbe et al., 2024). A lower Na_2SiO_3/NaOH ratio (i.e., <1) produces insufficient silicates, a weak gel, and low strength; a moderate Na_2SiO_3/NaOH ratio (1–2.5) leads to dense gel formation and good mechanical strength. Comparatively, a higher Na_2SiO_3/NaOH ratio (i.e., >2.5) leads to increased viscosity, poor workability, and risk of efflorescence. From the foregoing, it can be observed that higher hydroxide concentrations and silicate content generally reduce workability due to increased viscosity. Achieving balance is crucial to ensure sufficient working time without compromising strength. Generally, a combination of 10 M NaOH, a Na_2SiO_3/NaOH ratio of 1.5, and an activator/precursor ratio of 0.4 yields satisfactory optimal geopolymer strength for structural applications (Matsimbe et al., 2024).

2.3 MIXING PROCEDURE

The formation of a geopolymer is driven by a complex chemical reaction process known as geopolymerisation (Davidovits, 2020), which transforms aluminosilicate precursors such as fly ash, bottom ash, or GGBFS, into a hardened, cementitious binder through alkaline activation (Duxson et al., 2007; Provis and van Deventer, 2009). The resulting amorphous and/or semi-crystalline 3D network is responsible for the superior performance of the hardened geopolymer binders compared to ordinary cementitious systems (Elchalakani et al., 2023). Although the geopolymerisation process takes place continuously, it typically consists of four primary stages:

- *Dissolution:* Under highly alkaline conditions, aluminosilicate precursors dissolve, releasing silicate and aluminate species into the solution. The extent of dissolution determines the quantity of reactive monomers available to take part in polymerisation. High dissolution rates lead to denser gels and higher early strength, while poor dissolution (e.g., low amorphous content, coarse particles, or weak activators) results in slow reaction kinetics, porous microstructure, and lower early and long-term strength.
- *Transportation and orientation:* The dissolved species reorganize in the presence of alkali cations. Once dissolved, the silicate and aluminate species move through the activating medium by diffusion. They gradually reorganize, forming initial oligomers. This transitional phase is influenced by solution viscosity, temperature, water content, and ionic concentration (Na^+, K^+, Ca^{2+}). The efficient movement and collision of reactive species accelerate the nucleation of gel networks. High viscosity or insufficient water reduces mobility, slowing the gain of strength. The optimal diffusion of the reaction species leads to faster gel formation, more uniform microstructure, and mechanical continuity.
- *Polycondensation:* The silicate and aluminate species react to form a polymeric gel network of Si-O-Al bonds. The formation of N-A-S-H and/or

C-A-S-H gels provides the backbone of hardened binders. Polycondensation directly controls setting time, early mechanical strength, and dimensional stability. More rapid gel formation leads to faster early strength, while prolonged gel maturation contributes to long-term strength.

- *Hardening:* Upon completion of the gel network formation, continued reaction produces further cross-linking, a rigid matrix, and the gradual filling of pores. The process consists of gel maturation, residual dissolution, secondary precipitation, and reorganisation of structural units. Water not chemically bound either evaporates during curing or remains as physically bound water, contributing to porosity.

The geopolymer mixing procedure is a critical step that drives the geopolymerisation process and determines the fresh and hardened engineering properties of the final geopolymer product (Hardjito and Rangan, 2005). Proper selection and proportioning of materials, careful mixing, and effective control of curing are essential to achieve the desired performance characteristics. By adhering to a systematic mixing approach, geopolymer production can be optimised for structural applications, environmental sustainability, and industrial scalability. The mixing approach begins with the preparation of an activator solution. The NaOH or KOH solution is prepared by dissolving the solid pellets in distilled water to achieve the desired molarity (typically 8–14 M). The solution is allowed to cool for at least 24 hours to avoid heat-related issues during mixing. After cooling, the NaOH or KOH solution is mixed with Na_2SiO_3 solution in a predetermined ratio of Na_2SiO_3/NaOH, commonly between 1.0 and 2.5 (Matsimbe et al., 2024). The combined solution is stirred thoroughly and left to rest for at least 30 minutes before adding to the precursor material. Figure 2.5 illustrates the geopolymer preparation procedure.

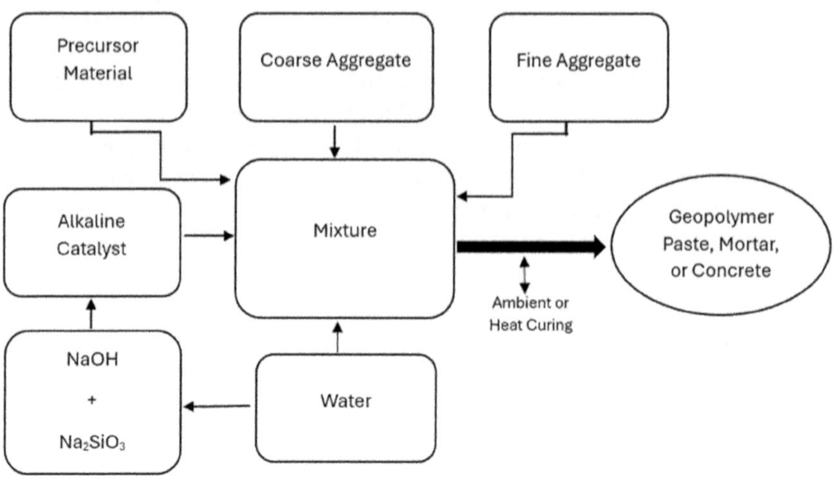

FIGURE 2.5 Preparation procedure for geopolymers.

Thereafter, the mixing approach proceeds by gradually adding the NaOH + Na_2SiO_3 solution to the dry aluminosilicate precursor material (e.g., fly ash, slag, etc.), while continuously stirring in a mechanical mixer to achieve homogeneity. The mixture is mixed thoroughly for 3–5 minutes until a uniform paste or mortar is formed. If additives are used (e.g., fibres, plasticizers, etc.), introduce them at this stage and mix for an additional 2 minutes. Carefully control the liquid-to-solid or activator-to-binder ratio (commonly between 0.3 and 0.6) to ensure suitable workability and strength (Hardjito and Rangan, 2005; Matsimbe et al., 2024). Fine aggregates (e.g., natural river sand), coarse aggregates (e.g., crushed granite/basalt), and fillers (e.g., micro-silica/limestone powder) form the bulk of geopolymer mortar and concrete. Aggregate quality should conform to relevant civil engineering standards, ensuring minimal impurities and adequate mechanical strength.

Immediately after mixing, cast the fresh geopolymer mixture into moulds. A vibratory table or manual tapping is used to eliminate air bubbles and ensure compaction. Avoid over-vibration to minimize segregation. Thereafter, the moulds are covered with plastic sheets to prevent moisture loss. The moulds are either heat-cured at 40–80°C for 8–48 hours in an oven or ambient-cured at 20–30°C for 7–28 days to enhance polymerisation and achieve strength. Both heat-cured and ambient-cured geopolymers show optimal mechanical performance required for structural applications (Provis and van Deventer, 2009; Elchalakani et al., 2023). However, heat-cured geopolymers gain strength more quickly and are suitable for precast applications, while ambient-cured geopolymers gain strength more slowly and are suitable for cast-in-situ applications, promoting SDG 13 (Matsimbe et al., 2024). Steam curing and microwave curing are alternative methods for achieving rapid strength gain in the precast industry. The curing temperature influences the geopolymerisation process because both thermodynamics (i.e., energy changes) and kinetics (i.e., reaction rates) govern the dissolution of precursors, gel formation, and microstructural development. Although elevated temperatures generally accelerate the setting and early strength gain, they can also promote microcracking due to moisture movement and thermal stresses.

For safety assurance, just like any other engineering activity, it is recommended to wear gloves, goggles, and protective clothing and to ensure proper ventilation during the mixing procedure. Although standardisation and scaling of geopolymers remain ongoing challenges, rapid advances in material science, processing technologies, and environmental regulations are accelerating their adoption in the construction industry.

PROBLEMS

1. Explain the main stages of geopolymerisation and the role of each stage in strength development.
2. Compare the chemical mechanisms of geopolymer and ordinary Portland cement hydration.

3. What are the main differences between using NaOH and KOH as alkaline activators?
4. A geopolymer mix uses 400 kg/m³ of fly ash and an activator-to-binder ratio of 0.45. If the activator solution consists of 60% sodium silicate and 40% sodium hydroxide, determine the quantities of each component required per cubic meter of concrete.
5. Describe the role of alkaline activators in geopolymer synthesis and how activator molarity affects strength development.
6. Why are slag-based geopolymers suitable for ambient curing, whereas fly ash–based geopolymers often require heat curing?
7. A geopolymer concrete uses a binder composed of 70% fly ash and 30% slag. Propose a curing regime suitable for this blend, and justify your choice based on reactivity and calcium content.
8. Using thermodynamic principles, explain why increasing curing temperature accelerates geopolymerisation but may also induce microcracking.
9. Design an experiment to evaluate the effect of activator molarity on setting time and compressive strength of fly ash–based geopolymer paste.
10. Propose a sustainable geopolymer mix incorporating phosphogypsum as a partial precursor replacement and justify its engineering performance.
11. A fly ash-phosphogypsum–based geopolymer mortar is prepared with an activator-to-binder ratio of 0.4 and NaOH molarity of 10 M. If 1 m³ of mortar requires 212 kg of binder, calculate the required mass of activator solution and anticipated compressive strength.

REFERENCES

ASTM C618. (2022). *Standard Specification for Coal Fly Ash and Raw or Calcined Natural Pozzolan for Use in Concrete.* Annual book of ASTM Standards, ASTM International, West Conshohocken, PA.

Davidovits, J. (2020). *Geopolymer Chemistry and Applications* (5th ed.). Institut Géopolymère, Saint-Quentin, France.

Duxson, P., Fernández-Jiménez, A., Provis, J. L., Lukey, G. C., Palomo, A., and van Deventer, J. S. J. (2007). Geopolymer technology: The current state of the art. *Journal of Materials Science,* vol. 42, no. 9, 2917–2933.

Elchalakani, M., Yang, B., Mao, K., and Pham, T. (2023). *Geopolymer Concrete Structures with Steel and FRP Reinforcements.* Woodhead Publishing Ltd, Kidlington, UK.

Hardjito, D., and Rangan, B.V. (2005). *Development and Properties of Low-Calcium Fly Ash-Based Geopolymer Concrete.* Curtin University of Technology, Kidlington, UK.

Matsimbe, J., Dinka, M., Olukanni, D., and Musonda, I. (2025). Characterization and valorization of fly ash and phosphogypsum into a binary composite for construction. *Cleaner Waste Systems*, vol. 12, no. 100355.

Matsimbe, J., Dinka, M., Olukanni, D., and Musonda, I. (2024). Performance evaluation and mix design of ambient-cured fly ash-phosphogypsum blended geopolymer paste and mortar. *Results in Engineering*, vol. 24, no. 103280.

Matsimbe, J., Dinka, M., Olukanni, D., and Musonda, I. (2023). Bibliometric trends of geopolymer research in Sub-Saharan Africa. *Materials Today Communications*, vol. 35.

Matsimbe, J., Dinka, M., Olukanni, D., and Musonda, I. (2022). Geopolymer: A systematic review of methodologies. *Materials*, vol. 15, no. 19.

Provis, J., and van Deventer, J. (2009). *Geopolymers: Structures, Processing, Properties and Industrial Applications*. Woodhead Publishing Ltd, Curtin University, Australia.

3 Geopolymer Properties Testing

3.1 PHYSICAL PROPERTIES

Fresh properties, such as workability and setting time, govern the ease of placement, compaction, and finishing. Soon after mixing the alkaline activator and precursor materials, the wet mixture is tested for workability and setting time. Workability reflects the ease of mixing, placing, and compacting geopolymer concrete or mortar. The slump test (ASTM C143/C143M, 2020) and flow table test (ASTM C230/C230M, 2021) are commonly employed. Factors affecting workability include activator concentration, solid/liquid (S/L) ratio, and type of precursor material (Duxson et al., 2007; Provis and van Deventer, 2009). Table 3.1 provides an integrated summary of the impact of activator concentrations, solid/liquid ratio, and precursor type on the workability of geopolymers. Increasing the concentration of NaOH or KOH increases the ionic strength and viscosity of the activator solution. This makes the fresh geopolymer mix thicker and stickier, reducing flowability. Higher molarity concentrations (i.e., ≥12 M) accelerate the dissolution of aluminosilicate particles, causing rapid formation of N-A-S-H or C-A-S-H gels, which eventually lead to rapid stiffening, shorter setting time, and poor workability. Lower molarity (6–10 M) improves flow but may reduce early strength due to slower dissolution. Higher silicate content increases viscosity because Na_2SiO_3 solutions are inherently viscous. Furthermore, workability depends heavily on the physical properties (particle size, shape) and chemical reactivity of the precursor.

Setting time defines the transition from a fluid to a rigid state and is influenced by the chemical reactivity of precursors and the molarity of the activator. The Vicat needle method (ASTM C191, 2021) is often adapted for geopolymers. Low-calcium fly ash results in longer setting times, while the inclusion of GGBFS or phosphogypsum accelerates setting due to the availability of calcium ions, which promote geopolymer gel formation (Matsimbe et al., 2024b). In low-calcium systems (e.g., Class F fly ash), geopolymerisation relies mainly on the dissolution of aluminosilicate glass and subsequent polycondensation to form N-A-S-H gel. This reaction is slower than the formation of C-A-S-H gel found in high-calcium

DOI: 10.1201/9781003787389-3

TABLE 3.1
Impact of geopolymer parameters on workability

Parameter	Effect on Workability	Possible Reasons
High activator concentration	↓ Decreases	Higher viscosity + rapid gel formation
High S/L ratio	↓ Decreases	Less liquid → more particle friction
Low S/L ratio	↑ Increases	More liquid → better flow
Fly ash	↑ Increases	Spherical particles, slow reaction
GGBFS	↓ Decreases	Angular particles, fast Ca-rich reaction
Bottom ash	↓ Decreases	Porous, irregular, absorbs liquid
Phosphogypsum	↓ Decreases	Accelerates setting

systems (e.g., Class C fly ash, PG, or GGBFS). With less calcium available, the system lacks fast-precipitating phases that normally accelerate stiffening.

Low-calcium fly ash (Class F) typically leads to longer setting times in geopolymer systems due to its limited calcium content, slower dissolution rate, and reduced nucleation capacity (Matsimbe et al., 2024b). Class F fly ash consists mainly of stable aluminosilicate glass with a high Si/Al ratio, which dissolves slowly in alkaline activators. As a result, the formation of the N-A-S-H geopolymer gel is delayed, resulting in longer dormant and induction periods during the early stages of the reaction. In contrast, high-calcium fly ash (Class C) contains reactive Ca-bearing phases that promote the rapid formation of C-A-S-H and hybrid gels (Provis and van Deventer, 2009). These phases dissolve more easily and provide nucleation sites that accelerate the structuring of the geopolymer matrix, resulting in significantly shorter setting times. Class C fly ash systems may even experience rapid stiffening or flash setting due to the high reactivity of calcium-rich components. Because of these differences, low-calcium geopolymer systems often require elevated temperatures to achieve practical setting times, whereas high-calcium geopolymer systems generally set adequately at ambient conditions.

Bulk density and specific gravity provide insight into material compactness and potential porosity (Matsimbe et al., 2025). Geopolymer concrete typically exhibits lower density (2000–2400 kg/m^3) and lower specific gravity than OPC concrete, primarily due to the nature of the precursor materials, reaction products, and microstructural characteristics of geopolymer binders. Measurements are carried out using ASTM C138/C138M (ASTM C138/C138M, 2024) for fresh concrete or ASTM C642 (ASTM C642, 2022) for hardened specimens. Table 3.2 provides the comparison between the density and specific gravity of geopolymer concrete and OPC concrete. The raw materials used in geopolymers, especially fly ash, bottom ash, phosphogypsum, and other industrial by-products, have lower specific gravity than ordinary Portland cement (Matsimbe et al., 2024b; Matsimbe et al., 2025). When low-density materials such as fly ash dominate the binder formulation,

TABLE 3.2
Comparison of density and specific gravity of OPC vs. geopolymer

Property	OPC Concrete	Geopolymer Concrete	Reason for the Difference
Specific gravity of binder	3.10–3.20	2.00–2.80	OPC clinker minerals (C_3S, C_2S, C_3A, C_4AF) are denser than aluminosilicate precursors (fly ash, bottom ash, phosphogypsum).
Specific gravity of aggregates	2.60–2.80	2.60–2.80	Aggregates are typically the same in both systems.
Bulk density of binder powder	830–1650 kg/m^3	400–1200 kg/m^3	Hollow/porous fly ash cenospheres and lower packing efficiency affect geopolymer bulk density.
Presence of dense crystalline phases	($Ca(OH)_2$, ettringite)	Mostly amorphous gel, minimal crystals	OPC hydration creates denser phases compared to geopolymers.
Effect of particle morphology	Cement: angular, dense	Fly ash/bottom ash: spherical, hollow/porous	Fly ash morphology reduces bulk density and improves workability.

the overall binder specific gravity decreases, resulting in a lower density in the hardened concrete.

Water absorption and porosity tests assess the extent of open pores, which influence durability (Matsimbe et al., 2024a). These tests, conducted in accordance with ASTM C642 (ASTM C642, 2022), involve drying, immersing, and reweighing specimens to determine water uptake. Water absorption and porosity in geopolymers are controlled by the reactivity and morphology of precursors, activator chemistry, mix design, and curing conditions. Low-calcium fly ash often yields higher porosity due to slower reaction rates, while high-calcium fly ash, GGBFS, and PG yield denser microstructures, reducing porosity and water absorption. Typical water absorption ranges from 3 to 15%, and porosity ranges from 8 to 30%, depending on the precursor materials and curing conditions (Matsimbe et al., 2024a). Lower absorption values (typically below 5%) indicate dense geopolymer matrices, attributed to the formation of gel structures depending on the precursor and mix design.

Drying shrinkage occurs as moisture evaporates during both curing and the service life of the material. ASTM C157/C157M (ASTM C157/C157M, 2021) procedures are adapted to measure the linear deformation. The factors influencing the shrinkage of geopolymers include curing temperature, type of activator, and the content of fly ash. Compared to OPC, geopolymers exhibit lower shrinkage due to reduced capillary stress; however, excessive activator solutions and high curing temperatures can increase the risk of shrinkage (Elchalakani et al., 2023). Geopolymers exhibit lower drying shrinkage than OPC primarily because their pore structure generates lower capillary stress during moisture loss. In OPC systems, evaporation from capillary pores forms water menisci that exert high suction forces, spearheading significant contraction of the C-S-H gel. In contrast, geopolymers have fewer capillary pores, a larger proportion of stable gel pores, and a highly cross-linked aluminosilicate network that is intrinsically more resistant to deformation. Additionally, water in geopolymers is usually a processing medium rather than a chemically bound component, so its removal does not collapse the gel matrix (Matsimbe et al., 2024b). These factors collectively reduce capillary tension and shrinkage deformation in geopolymer concrete.

3.2 MECHANICAL PROPERTIES

Mechanical properties define the hardened engineering capacity and structural load-bearing capability of geopolymers. Typically, compressive strength is the most modelled engineering property as it directly influences structural performance and safety rating (Matsimbe et al., 2022; Matsimbe et al., 2025). It is determined using a universal testing machine in accordance with ASTM C109 (ASTM C109/C109M, 2020). The compressive strength test is usually done on 50 mm geopolymer cubes or 100 mm geopolymer cylinders after heat curing for 24 hours or ambient curing for 28 days. The compressive strength values of geopolymers usually range from 20 MPa to 80 MPa, depending on precursor type, curing temperature, and activator concentration (Matsimbe et al., 2024b). Elevated curing (60–90°C) enhances early strength, particularly in fly ash–based geopolymer systems. The lower end strength values (e.g., 20 MPa) are typical for ambient-cured, low-reactivity precursors (Class F fly ash, bottom ash) at early curing ages (1–7 days). The upper-end strength values (e.g., 40 MPa) are achievable with highly reactive precursors (GGBFS, Class C fly ash, PG) combined with optimized activator chemistry, good particle packing, and efficient curing. However, it should be noted that strength is not fixed but rather varies depending on precursor chemistry, activator formulation, water/activator content, S/L ratio, curing regime, and age.

Furthermore, higher amorphous SiO_2/Al_2O_3 (Si/Al) content in precursors increases the pool of reactive species available for geopolymerisation and subsequent strength development (Matsimbe et al., 2022). An optimum Si/Al (often 2–3 for fly ash–based systems) produces a cross-linked N-A-S-H network with good strength properties. High-calcium content (CaO) precursors (GGBFS, Class C fly ash, phosphogypsum) promote the formation of C-A-S-H or hybrid C-A-S-H/N-A-S-H gels, which densify the matrix and speed early strength gain. Low CaO precursors rely mainly on N-A-S-H gel, which can be mechanically excellent but forms more slowly. The presence of finer particles, which provide a higher specific surface area, promotes faster dissolution and produces more homogeneous gels with better packing, facilitating the development of strength. Spherical fly ash improves packing and workability, but, if too coarse, it slows reaction and lowers strength. The presence of impurities and unreactive crystalline phases (quartz, mullite) reduces available reactive mass and lowers achievable compressive strength.

Elevated temperature accelerates the rate at which Si and Al are leached from the precursor glass, raising the concentration of reactive species in solution. The reaction kinetics (oligomerization → gel growth → network formation) accelerate significantly with temperature, so a cohesive load-bearing gel network develops faster (Provis and van Deventer, 2009). Heat curing reduces capillary porosity by promoting more complete reaction and better packing of reaction products, resulting in higher early compressive strength. Most of the fly ash–based geopolymers that are weak at ambient curing reach target strengths within 24–48 hours when heat-cured at 60–80°C. However, it is common knowledge that excessive temperature or too-rapid drying can cause differential shrinkage, microcracking, or efflorescence. Typical laboratory practice uses temperature ranges of 60–80°C for 6–24 hours for optimal mechanical properties, followed by ambient curing to reach 28 days in duration.

Flexural strength, often referred to as the modulus of rupture, reflects a material's ability to resist bending or transverse loading, determined in accordance with ASTM C78 (ASTM C78, 2010) or ASTM C293 (ASTM C293, 2010). It is a key indicator of the geopolymer's tensile performance and overall structural integrity, especially in applications such as beams, slabs, and pavement blocks where bending stresses are dominant. Geopolymer mortar and concrete exhibit flexural strengths ranging from 4 to 10 MPa, which is typically 10–20% of compressive strength (Matsimbe et al., 2024b). The fracture surfaces often show a more cohesive failure pattern due to the strong interfacial transition zone (ITZ) between binder and aggregate. Incorporating slag or fine silica powder enhances the flexural behaviour of geopolymers by improving ITZ and matrix homogeneity. The flexural strength (σ_x), for centre-point loading, is calculated using the formula:

$$\sigma_x = \frac{3PL}{2bd^2} \tag{3.1}$$

where P is the maximum load (N), L is the span length (mm), b is the specimen width (mm), and d is the specimen depth (mm).

Splitting tensile strength, measured using ASTM C496 (ASTM C496/C496M, 2017), determines the tensile capacity of geopolymer concrete. It is crucial for assessing the resistance of geopolymers to cracking, shrinkage-induced stresses, and flexural loading. Since geopolymers, like many other cementitious composites, are brittle and weak in tension, the splitting tensile strength provides valuable insights into their tensile performance and bonding quality between binder and aggregate. Its strength values generally range between 2 and 5 MPa, depending on binder type, activator concentration, and curing regime. The presence of calcium-bearing materials, such as GGBFS, improves tensile performance through the co-formation of gel structures. The splitting tensile strength (σ_t) is determined as:

$$\sigma_t = \frac{2P}{\pi LD} \tag{3.2}$$

where P is the maximum load at failure, L is the length of the specimen, and D is the diameter of the specimen.

3.3 MICROSTRUCTURAL PROPERTIES

Microstructural characterisation provides fundamental insights into the chemical composition, reaction mechanisms, and phase development of geopolymers. These analyses are essential for understanding how the chemistry of precursor materials, alkaline activator concentration, and curing conditions influence the structure–property relationships of geopolymer binders (Provis and van Deventer, 2009; Matsimbe et al., 2022). It provides a scientific foundation for correlating material composition and processing with mechanical and durability performance. The key techniques consist of X-ray fluorescence (XRF), scanning electron microscopy with energy-dispersive spectroscopy (SEM-EDS), X-ray diffraction (XRD), Fourier transform infrared spectroscopy (FTIR), and thermogravimetric analysis (TGA). The combination of XRF, SEM-EDS, XRD, FTIR, and TGA enables a comprehensive understanding of the reaction chemistry, phase composition, and morphological evolution of geopolymer systems, guiding optimisation for strength, durability, and sustainability.

XRF is a primary analytical technique used to quantify the elemental and oxide composition of geopolymer precursors (e.g., fly ash, GGBFS, phosphogypsum, bottom ash [see Table 3.3]) and hardened geopolymer matrices. Because

TABLE 3.3
Chemical composition of bottom ash (BA)

Constituent	% mass	Constituent	% mass
SiO_2	52.02	TiO_2	1.53
Al_2O_3	24.41	MnO	0.06
Fe_2O_3	8.73	BaO	0.13
CaO	5.98	NiO	–
MgO	1.34	V_2O_5	–
Na_2O	-	Cr_2O_3	–
K_2O	0.55	LOI	4.22
SO_3	0.30		
P_2O_5	0.73		

geopolymerisation is fundamentally controlled by precursor chemistry, particularly the availability of reactive silica (SiO_2) and alumina (Al_2O_3), XRF provides essential data for predicting reactivity, tailoring activator formulations, and understanding mechanical and durability performance.

The crucial aspects served by XRF analysis comprise determining precursor suitability by quantifying SiO_2, Al_2O_3, and CaO contents; estimating network-forming and network-modifying species that influence gel structure (N-A-S-H, C-A-S-H, or hybrid gels); identifying impurities (e.g., Fe_2O_3, TiO_2, SO_3) that may delay or inhibit dissolution; calculating oxide ratios such as Si/Al, Ca/Si, and alkali content, which are central parameters for geopolymer chemistry; classifying materials (e.g., ASTM C618 Class F vs. Class C fly ash); and monitoring bulk chemistry of the hardened geopolymer to verify reaction completeness and presence of unreacted precursors (Provis and van Deventer, 2009; Matsimbe et al., 2024b).

SEM-EDS is crucial for visualizing reaction products such as N-A-S-H or C-A-S-H gels; identifying unreacted precursor particles (fly ash spheres, slag grains, gypsum crystals); assessing reaction extent and degree of dissolution; understanding pore structure; cracking behaviour and microdefects; determining elemental ratios (Si/Al, Ca/Si, Na/Al) at the microscale; mapping elemental distribution across gel phases and inclusions; and correlating microstructure with mechanical and durability performance (Matsimbe et al., 2024b). Figure 3.1 shows SEM-EDS equipment installed at the laboratory and used for microstructural analysis.

SEM provides morphological details, revealing matrix densification and the presence of unreacted particles, while an EDS offers elemental mapping, confirming

FIGURE 3.1 SEM-EDS equipment.

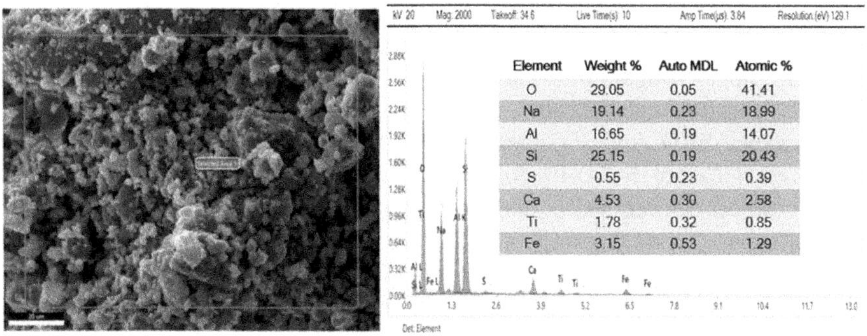

FIGURE 3.2 SEM micrograph with EDS analysis for a low calcium FA-based geopolymer paste.

TABLE 3.4
Linking SEM-EDS to geopolymer performance

Microstructural feature	Property implication
Dense, cohesive gel	High compressive strength
Many unreacted particles	Reduced reactivity, weaker matrix
Large pores and voids	Higher permeability, lower durability
Well-distributed Ca	Improved early strength (hybrid gels)
Microcracks	Shrinkage issues, lower durability
S-rich PG inclusions	Improved strength, possible expansion, or setting issues

the presence of Si, Al, Ca, and Na. As observed in Figure 3.2, SEM-EDS helps in understanding the morphology, crack intensity, reacted/unreacted particles, and gel distribution in geopolymers.

The dominant binding phase in low-calcium geopolymer systems is N-A-S-H gel, which appears as an amorphous, dense matrix with smooth to slightly textured regions, and lacks the plate-like or layered morphology characteristic of C-S-H. When GGBFS, Class C fly ash, or PG is present in the geopolymer system, C-A-S-H or N-C-A-S-H hybrid gels form, which appear denser and more homogeneous with fibrous or foil-like textures partially layered due to Ca incorporation (Provis and van Deventer, 2009; Matsimbe et al., 2022). Fly ash-based geopolymers often show spherical unreacted particles, while slag-based systems display a denser, amorphous structure. Table 3.4 provides a summary of the connection between SEM-EDS and geopolymer performance, guiding the practicing engineer in understanding the fundamental principles influencing the physico-mechanical properties of geopolymers.

XRD is a fundamental characterisation technique used to identify crystalline phases and evaluate amorphous content in geopolymer precursor materials and hardened binders (Elchalakani et al., 2023). Because geopolymerisation is predominantly an amorphous aluminosilicate polymerisation process, XRD provides critical insights into the initial mineralogical composition, dissolution potential, reaction mechanisms, and the degree of amorphous gel formation. The Rietveld XRD analysis, supported by the Powder Diffraction File of the International Centre for Diffraction Data and OriginLab software, is used to quantify the amorphous and crystalline phases. The main indicator of geopolymer gel formation is a broad amorphous hump around 20–35° 2$\boldsymbol{\theta}$. A high amorphous content (>60%) indicates high reactivity and suitability for geopolymerisation. The hump shifts slightly to lower angles due to polymerisation and increased structural disorder, and intensity changes reflect the degree of gel formation and unreacted material. This amorphous hump corresponds to N-A-S-H, C-A-S-H, or hybrid gels. In fly ash or bottom ash geopolymers, broad humps at 20–35° 2$\boldsymbol{\theta}$ indicate amorphous geopolymer gel formation. Peaks corresponding to quartz, mullite, or hematite (FA), merwinite or akermanite (GGBFS), and gypsum (PG), suggest residual unreacted material. Depending on the chemistry and curing conditions, new crystalline phases, including zeolites (e.g., sodalite, faujasite, zeolite A or P), may form (Provis and van Deventer, 2009; Matsimbe et al., 2022). Overall, XRD provides invaluable insights into geopolymer mix design, reaction extent, and long-term durability.

FTIR is a rapid, non-destructive technique that probes vibrational modes of chemical bonds and functional groups. In geopolymer science, FTIR is widely used to monitor the evolution of chemical bonds during dissolution-polycondensation, to identify secondary reaction products, and to infer the degree and nature of gel formation (Provis and van Deventer, 2009). The method complements XRD (which detects crystalline or amorphous phases) and SEM-EDS (which maps morphology and elemental composition) by providing molecular-level bonding information from both raw precursors and hardened binders. FTIR analysis provides a direct probe of the evolving aluminosilicate network during the geopolymerisation process (Matsimbe et al., 2022). The diagnostic signature is the shift of the main Si-O-T asymmetric stretching band from ~1080–1100 cm^{-1} (glass phase in precursors) to ~950–980 cm^{-1} in the geopolymer binder. This shift, together with band broadening and diminution of water-related absorptions (~3400 and 1640 cm^{-1}), reflects the dissolution of glassy particles and subsequent polycondensation into an amorphous, cross-linked N-A-S-H (or hybrid) gel. The careful deconvolution of the 1200–700 cm^{-1} region, supported by XRD and SEM-EDS, yields quantitative insight into reaction progress and the influence of Si/Al, Ca content, activator composition, and curing regime on network structure.

TGA is a thermal analytical technique that measures the mass change of a sample as it is heated under a controlled atmosphere. It quantifies free and bound water, dehydroxylation, carbonate, or sulphate content, and organic/unburnt carbon (Provis and van Deventer, 2009). Furthermore, it assesses thermal stability and the temperatures at which structural decomposition occurs. Coupled with derivative curves and evolved-gas analysis, TGA gives powerful mechanistic insight into geopolymer composition, degree of reaction, and high-temperature behaviour. It evaluates thermal stability and weight loss below 200°C due to physically bound water and minor losses around 800°C related to structural decomposition and high-temperature stability. This is achieved by measuring the mass change of a geopolymer sample as a function of temperature or time under a controlled atmosphere. Ca-rich systems exhibit improved thermal resistance due to denser gel structures.

3.4 DURABILITY PROPERTIES

Durability evaluates the long-term performance and resistance of geopolymers to environmental and chemical attacks that can degrade their structure and mechanical integrity. Compared to OPC, geopolymers often exhibit comparable or superior resistance to aggressive environments due to their dense, low-calcium, and highly cross-linked aluminosilicate network (Provis and van Deventer, 2009; Elchalakani et al., 2023).

The water absorption and porosity of a geopolymer provide valuable insights into its ability to absorb and retain water, which influences its resistance to aggressive environments. Water absorption and porosity are determined in accordance with ASTM C1403 (ASTM C1403, 2024) and ASTM C642 (ASTM C642, 2022), comprising the oven-drying, water immersion, and boiling stages. Compared to OPC concrete, geopolymer composites exhibit lower water absorption (2–8%) and porosity (5–15%) due to their compact microstructure (Matsimbe et al., 2024a). Lower water absorption and porosity indicate a denser matrix, improved impermeability, and greater chemical durability.

Sulphate attack occurs when sulphates, present in the environment, react with the geopolymer paste, leading to expansion, cracking, and ultimately the disintegration of the geopolymer. Sulphate attack resistance in geopolymers is determined in accordance with ASTM C1012 (ASTM C1012/C1012M, 2024), which requires immersion in Na_2SO_4 or $MgSO_4$ solutions for several months and measuring changes in mass, expansion, and residual strength (Matsimbe et al., 2024a). Geopolymers, especially low-calcium fly ash–based systems, exhibit excellent resistance due to the absence of portlandite ($Ca(OH)_2$), which typically reacts with sulphates in OPC. The resistance to sulphate attack indicates the suitability of constructed structures in sewage, saline soils, and marine conditions.

Chloride attack in geopolymers primarily involves the ingress of chloride ions, which can lead to the corrosion of steel reinforcement in reinforced geopolymer concrete structures. This corrosion can significantly reduce the strength, serviceability, and longevity of the structure. Chloride attack is determined using the rapid chloride permeability test (ASTM C1202 [ASTM C1202, 2025]). Geopolymers typically exhibit low charge passed (<1000 Coulombs), indicating high resistance to chloride ingress in coastal or marine conditions. While geopolymer concrete generally exhibits good resistance to chloride penetration, variations in precursor materials, mix design, and curing conditions can influence its behaviour in chloride-rich environments.

Acid resistance assesses the chemical stability of geopolymers in acidic environments. Geopolymers can be vulnerable to strong acids such as sulfuric acid (H_2SO_4) and hydrochloric acid (HCl). The primary acid attack mechanism involves the breakdown of the geopolymer's aluminosilicate structure through ion exchange and electrophilic attack, leading to reduced strength and degradation. While geopolymer formulations exhibit better resistance to acid attack than OPC, they can still experience deterioration under prolonged acid exposure and higher calcium content formulations. Immersion in acid solutions for 28–90 days leads to changes in weight loss, strength retention, and surface integrity (Matsimbe et al., 2024a). Fly ash–based geopolymers show mass loss below 10% after 28 days, outperforming OPC concrete due to their aluminosilicate backbone and absence of ($Ca(OH)_2$), which resists acid dissolution.

Carbonation resistance measures the depth and rate of CO_2 penetration that can alter pH and destabilise the geopolymer matrix. It is crucial for geopolymer durability, as it affects the materials' ability to protect embedded steel reinforcement from corrosion (Provis and van Deventer, 2009). The primary attack mechanism involves the diffusion of CO_2 into the geopolymer matrix and reacting with the geopolymer's reaction products (C-[A]-S-H gel) to form carbonates. This reaction can initially fill pores and reduce permeability, but continued carbonation can lead to a decrease in the volume of C-[A]-S-H gel, increasing porosity and weakening the material. A performance-based approach is recommended for developing standards, as two geopolymer concretes with similar mix designs can exhibit different carbonation behaviours. ASTM C1157 (ASTM C1157, 2010) is often used as a reference for evaluating the performance of geopolymer concrete against carbonation attack. The method involves conducting accelerated carbonation tests under controlled CO_2 concentration, temperature, and humidity, followed by phenolphthalein indicator testing. Geopolymers generally exhibit lower carbonation resistance compared to OPC due to the absence of calcium hydroxide ($Ca(OH)_2$), which acts as a buffer. Low-calcium geopolymers show limited carbonation depth, whereas calcium-rich ones are more susceptible due to C-A-S-H carbonation reactions forming $CaCO_3$.

Furthermore, geopolymers exhibit excellent fire and thermal resistance, maintaining stability at high temperatures due to their inorganic nature and unique microstructure. They demonstrate reduced cracking, spalling, and minimal strength loss compared to OPC when exposed to heat. Geopolymers can withstand temperatures exceeding 1000°C, making them suitable for high-temperature applications and fire-resistant construction (Matsimbe et al., 2022). ASTM E119 (ASTM E119, 2024), ASTM C1113M (ASTM C1113/C1113M, 2024), and ASTM E228 (ASTM E228, 2023) are followed to determine the fire and thermal resistance of geopolymers. High-temperature resistance is a hallmark of geopolymers. After heating to temperatures of 600–900°C, geopolymers retain significant strength due to their ceramic-like properties. Thermal gravimetric analysis and residual strength tests confirm minimal degradation, in contrast to OPC, which experiences severe cracking from dehydration.

Alkali-silica reaction (ASR) is a detrimental chemical reaction that occurs in concrete when reactive silica in aggregates reacts with the highly alkaline pore solution of the cement, leading to expansion and cracking. While no specific ASTM standard has been developed exclusively for evaluating ASR resistance in geopolymers, the accelerated mortar bar test (ASTM C1260 [ASTM C1260, 2023]) and the concrete prism test (ASTM C1293 [ASTM C1293, 2023]) are commonly used to assess ASR behaviour in geopolymer concrete. While geopolymers contain alkalis, their resistance to ASR is often superior to that of OPC concrete due to the differences in pore solution chemistry and reaction mechanisms, which promote the consumption of alkalis during the geopolymerisation process. The type of aluminosilicate precursor and composition of the activator solution used in the geopolymer mix design are key factors influencing its ASR resistance.

Freeze-thaw resistance refers to a geopolymer's ability to withstand damage from the repeated freezing and thawing of water within its pores. This is especially important for geopolymer concrete exposed to cold climates, where repeated cycles of freezing and thawing can lead to scaling, cracking, and crumbling, ultimately compromising strength. Geopolymers exhibit comparable freeze-thaw resistance to OPC due to low permeability and flexible gel structures that accommodate internal stresses. The key factors influencing geopolymers' ability to resist freeze-thaw comprise the type of precursor material, alkaline activator type and ratio, liquid-to-binder ratio, and curing conditions. ASTM C666 (ASTM C666, 2017) is a common standard used for assessing concrete's resistance to freezing and thawing cycles.

PROBLEMS

1. Explain the importance of workability in geopolymer concrete, and describe one method to improve it.
2. Compare the typical setting times of fly ash–based and slag-based geopolymers, and explain the difference.
3. Describe the procedure for testing compressive strength of geopolymer mortar as per ASTM C109 (ASTM C109/C109M, 2020).
4. Using ASTM C642 (ASTM C642, 2022), calculate the porosity of a geopolymer specimen with the following data:
   - Oven-dry mass = 1950 g
   - Saturated mass = 2050 g
   - Submerged mass = 1850 g
5. Compare the ASTM C78 (third-point loading) and ASTM C293 (centre-point loading) methods in terms of stress distribution and failure behaviour.
6. Four geopolymer concrete samples (A, B, C, and D) have given strengths after 28 days of curing. Using the data in Table 3.5, develop a linear regression between flexural strength (σ_x) and compressive strength (σ_c) for a series of geopolymer mortars.

TABLE 3.5
Raw data

Sample	σ_x (MPa)	σ_c (MPa)
A	3.8	20
B	5.2	35
C	6.1	45
D	7.0	55

7. A cylindrical geopolymer concrete specimen with a diameter of 100 mm and a length of 200 mm fails under a maximum load of 52 kN during a splitting tensile strength test. Calculate the splitting tensile strength (σ_t) in MPa.
8. An XRF analysis of a fly ash sample yields the following oxide composition (in wt.%): SiO_2 = 52.3, Al_2O_3 = 27.4, Fe_2O_3 = 7.6, CaO = 6.2, MgO = 1.5, Na_2O = 0.8, K_2O = 1.2.
   - Calculate the Si/Al ratio.
   - Comment on the suitability of this material for geopolymer synthesis in terms of reactivity and strength potential.
9. In an XRD test, a geopolymer paste shows a broad hump between 22 and 35° 2$\boldsymbol{\theta}$, with residual crystalline peaks for quartz (SiO_2) and mullite ($3Al_2O_3{\cdot}2SiO_2$).
   - Explain what the broad hump represents.
   - Suggest how heat curing or the inclusion of GGBFS could alter this XRD pattern.
10. Explain how the shift in the FTIR band from ~1080 to ~950 cm^{-1} provides evidence of geopolymerisation.
11. Discuss how TGA and FTIR can be jointly used to evaluate the thermal performance of geopolymer binders.
12. Discuss two reasons why geopolymers generally resist sulphate attack better than OPC concrete.
13. Explain why some high-calcium geopolymers (e.g., those incorporating GGBFS) may have reduced carbonation resistance despite superior strength development.
14. Explain why geopolymers typically exhibit minimal alkali-silica reaction (ASR) even though they contain high alkali contents in their activators.
15. A fly ash–based geopolymer specimen is immersed in a 5% Na_2SO_4 solution. The 28-day compressive strength before immersion is 48 MPa, and after 180 days of exposure, it drops to 44 MPa.
   - Calculate the percentage strength retention.
   - Interpret whether the material demonstrates good sulphate resistance based on a minimum acceptable retention of 85%.

16. A geopolymer wall panel is exposed to an accelerated carbonation chamber (4% CO_2 concentration) for 28 days. The average carbonation depth measured with phenolphthalein is 2.5 mm. If the same mix was expected to have a carbonation coefficient of 5 mm/√year, estimate the carbonation depth after 5 years of natural exposure, assuming square-root-time dependence

REFERENCES

ASTM C78. (2010). *Standard Test Method for Flexural Strength of Concrete (Using Simple Beam with Third-Point Loading)*. ASTM International, West Conshohocken, PA, USA.

ASTM C191. (2021). *Standard Test Methods for Time of Setting of Hydraulic Cement by Vicat Needle*. ASTM International, West Conshohocken, PA, USA.

ASTM C293. (2010). *Standard Test Method for Flexural Strength of Concrete (Using Simple Beam with Center-Point Loading)*. ASTM International, West Conshohocken, PA, USA.

ASTM C642. (2022). *Standard Test Method for Density, Absorption, and Voids in Hardened Concrete*. ASTM International, West Conshohocken, PA, USA.

ASTM C666. (2017). *Standard Test Method for Resistance of Concrete to Rapid Freezing and Thawing*. ASTM International, West Conshohocken, PA, USA.

ASTM C1157. (2010). *Standard Performance Specification for Hydraulic Cement*. ASTM International, West Conshohocken, PA, USA.

ASTM C1202. (2025). *Standard Test Method for Electrical Indication of Concrete's Ability to Resist Chloride Ion Penetration*. ASTM International, West Conshohocken, PA, USA.

ASTM C1260. (2023). *Standard Test Method for Potential Alkali Reactivity of Aggregates (Mortar-Bar Method)*. ASTM International, West Conshohocken, PA, USA.

ASTM C1293. (2023). *Standard Test Method for Determination of Length Change of Concrete Due to Alkali-Silica Reaction*. ASTM International, West Conshohocken, PA, USA.

ASTM C1403. (2024). *Standard Test Method for Rate of Water Absorption of Masonry Mortars*. ASTM International, West Conshohocken, PA, USA.

ASTM C109/C109M. (2020). *Compressive Strength of Hydraulic Cement Mortars*. ASTM International, West Conshohocken, PA, USA.

ASTM C1012/C1012M. (2024). *Length Change of Hydraulic-Cement Mortars Exposed to a Sulfate Solution*. ASTM International, West Conshohocken, PA, USA.

ASTM C1113/C1113M. (2024). *Standard Test Method for Thermal Conductivity of Refractories by Hot Wire (Platinum Resistance Thermometer Technique)*. ASTM International, West Conshohocken, PA, USA.

ASTM C138/C138M. (2021). *Standard Test Method for Density (Unit Weight), Yield, and Air Content (Gravimetric) of Concrete*. ASTM International, West Conshohocken, PA, USA.

ASTM C143/C143M. (2020). *Standard Test Method for Slump of Hydraulic-Cement Concrete*. ASTM International, West Conshohocken, PA, USA.

ASTM C157/C157M. (2021). *Standard Test Method for Length Change of Hardened Hydraulic-Cement Mortar and Concrete*. ASTM International, West Conshohocken, PA, USA.

ASTM C230/C230M. (2021). *Standard Specification for Flow Table for Use in Tests of Hydraulic Cement*. ASTM International, West Conshohocken, PA, USA.

ASTM C496/C496M. (2017). *Standard Test Method for Splitting Tensile Strength of Cylindrical Concrete Specimens*. ASTM International, West Conshohocken, PA, USA.

ASTM E119. (2024). *Standard Test Methods for Fire Tests of Building Construction and Materials*. ASTM International, West Conshohocken, PA, USA.

ASTM E228. (2023). *Standard Test Method for Linear Thermal Expansion of Solid Materials with a Push-Rod Dilatometer*. ASTM International, West Conshohocken, PA, USA.

Duxson, P., Provis, J. L., Lukey, G. C., & van Deventer, J. S. J. (2007). The role of inorganic polymer technology in the development of green concrete. *Cement and Concrete Research*, vol. 37, no. 12, 1590–1597.

Elchalakani, M., Yang, B., Mao, K., and Pham, T. (2023). *Geopolymer Concrete Structures with Steel and FRP Reinforcements*. Woodhead Publishing Ltd, Kidlington, UK.

Matsimbe, J., Dinka, M., Olukanni, D., and Musonda, I. (2022). Geopolymer: A systematic review of methodologies. *Materials*, vol. 15, no. 19.

Matsimbe, J., Dinka, M., Olukanni, D., and Musonda, I. (2024a). Durability properties of ambient-cured fly ash-phosphogypsum blended geopolymer mortar in terms of water absorption, porosity, and sulfate resistance. *Discover Sustainability*, vol. 5, no. 330.

Matsimbe, J., Dinka, M., Olukanni, D., and Musonda, I. (2024b). Performance evaluation and mix design of ambient-cured fly ash-phosphogypsum blended geopolymer paste and mortar. *Results in Engineering*, vol. 24, no. 103280.

Matsimbe, J., Dinka, M., Olukanni, D., and Musonda, I. (2025). Characterization and valorization of fly ash and phosphogypsum into a binary composite for construction. *Cleaner Waste Systems*, vol. 12, no. 100355.

Provis, J., and van Deventer, J. (2009). *Geopolymers: Structures, Processing, Properties and Industrial Applications*. Woodhead Publishing Ltd, Cambridge. UK.

4 Geopolymer Design Modelling

4.1 TRIAL-AND-ERROR APPROACH

The trial-and-error approach, despite its perceived simplicity, flexibility, and accessibility, remains a foundational methodology in the development of geopolymers (Matsimbe et al., 2022). Especially valuable in early-stage investigations, it enables researchers to systematically explore material behaviour, refine formulations, and build an intuitive understanding. A practical framework for conducting trial-and-error mix design involves initial mix selection, parameter variation, preparation and casting, testing and evaluation, and iterative refinement until performance targets are met (Matsimbe et al., 2024b). The key parameters commonly varied during trial-and-error experimentation comprise precursor type and ratio, alkaline activator composition, liquid-to-solid ratio, curing conditions, and additives if necessary. While it may not replace data-driven or AI-based methods, it complements them, serving as a practical, low-cost, and inclusive approach in both well-equipped laboratories and resource-constrained environments. Despite being time-consuming, resource-intensive, and lacking predictive capability, trial-and-error remains useful for exploratory studies or when dealing with new raw materials, as it provides an intuitive understanding of material behaviour. Emphasis is placed on systematic iteration, practical adjustments, and field-based observations that inform the optimisation of strength, workability, setting time, and durability of geopolymers.

4.2 TAGUCHI APPROACH

The development of a geopolymer mix design is more intricate than conventional ordinary Portland cement systems due to the range of available precursors, the diversity of alkali activators, and the sensitivity of performance to curing conditions. Traditional methods such as trial-and-error are time-consuming and may not reveal interactions between variables (Matsimbe et al., 2022). The Taguchi method, a statistical design of experiments (DOE) approach, provides a structured pathway to identify optimal process parameters with reduced experimental effort. The Taguchi approach utilizes orthogonal arrays (OA) and signal-to-noise (S/N) ratios to

DOI: 10.1201/9781003787389-4

determine the most influential factors in a process and achieve robust performance. The OA facilitates studying many variables with a minimal number of experiments, while the S/N measures the robustness and consistency of performance, reducing variability caused by uncontrollable factors.

Depending on the target of the response variable (e.g., compressive strength), different S/N ratios are used to maximize strength or durability (larger is better) and minimize shrinkage or permeability (smaller is better). The application of the Taguchi approach in geopolymer mix design involves identifying performance objective, selecting input factors and levels, designing the OA based on the number of factors and levels (e.g., 4 factors at 3 levels can be studied using L9 (3^4) OA with just 9 experiments instead of 81), conducting experiments as per the OA, computing the S/N ratio for each trial based on the desired response, determining the factor level combinations that yield the highest S/N ratio, conducting the main effects plot, interaction plot, and analysis of variance (ANOVA) to identify statistically significant factors, and running a confirmation test using the optimal factor levels to validate the predicted performance.

4.3 RESPONSE SURFACE METHODOLOGY

Response surface methodology (RSM) is a powerful mathematical and statistical technique used to model and optimise geopolymer mix designs. By analysing the relationships between input variables and desired output, RSM helps identify the optimal mix proportions for desired properties (Matsimbe et al., 2022). This approach is more efficient than traditional methods, as it systematically explores the design space, often requiring fewer experiments than conventional methods. The RSM involves the identification of key independent variables/factors (e.g., raw material ratios, alkaline activator concentration, activator-to-binder ratio, silica modulus, water content, and curing temperature/duration) influencing the geopolymer response variable/outcome (e.g., strength, workability, setting time, water absorption, and shrinkage). RSM commonly utilizes the central composite design (CCD) and Box–Behnken Design (BBD) as DOE techniques to explore the parameter space. CCD is a popular choice for fitting quadratic models, allowing for curvature in the response surface involving factorial, axial, centre, contour, and 3D points. BBD requires fewer experiments and is useful when some factor combinations are undesirable.

RSM uses regression analysis to fit mathematical models (usually polynomial equations) to the experimental data. First-order models and second-order models (quadratic) capture interactions and curvature between factors, providing a more accurate representation of the response surface. ANOVA is used to identify which variables in the fitted model have a significant impact on the output. P-value (<0.05) indicates significant variables, F-value indicates model fit strength, and R^2 or adjusted R^2 show how well the model explains variability. The predicted optimal mix is then tested experimentally to validate the accuracy of the RSM model and the optimised outcomes. A significant agreement between the predicted and actual performance outcomes confirms model reliability. Furthermore, RSM can be coupled with machine learning models to enhance prediction accuracy and broaden the scope of mix design optimisation. The commonly used statistical software used in geopolymer design modelling consists of Design Expert, Minitab, SPSS, MATLAB/R/Python, and STATISTICA.

4.4 AI AND MACHINE LEARNING APPROACH

AI is an overarching field that encompasses various computational techniques designed to mimic human intelligence. ML, a subfield of AI, involves training algorithms on data to make predictions or decisions (Matsimbe et al., 2024a). AI and ML are being increasingly utilised to optimise geopolymers, particularly in predicting their mechanical properties and mix design. These techniques provide a powerful approach to analysing complex datasets and creating predictive models, thereby reducing the need for extensive experimental testing and potentially leading to more efficient and sustainable geopolymer production. AI and ML find extensive applications in the following areas:

4.4.1 Predicting the Mechanical Properties of Geopolymers

Supervised machine learning algorithms, e.g., linear regression (LR), Gaussian process regression (GPR), support vector machines (SVM), K-nearest neighbours (KNN), naive Bayes (NB), decision trees (DT), random forest (RF), and artificial neural networks (ANN), are used to predict the compressive strength of geopolymers based on their varying input parameters such as type and quantity of precursor, activator concentration, liquid-to-precursor ratio, and curing time and temperature, significantly reducing the need for trial-and-error experimentation.

LR is a supervised machine learning algorithm used to model the relationship between one or more input variables (independent variables) and an output variable (dependent variable). In the context of geopolymer design, the algorithm can be used to predict material properties, e.g., compressive strength, density, workability, or durability, based on mix design parameters and curing conditions. Mathematically, the general LR model is expressed as:

$$y = \beta_0 + \beta_1 x_1 + \beta_2 x_2 + \ldots + \beta_n x_n + \varepsilon \tag{4.1}$$

where:

- y is the predicted response (e.g., compressive strength, flexural strength, workability, setting time);
- $x_1, x_2, \ldots, x_n$ represent the independent input variables (e.g., fly ash content, phosphogypsum content, activator ratio, curing temperature);
- $\beta_0, \beta_1, \ldots, \beta_n$ are the regression coefficients;
- ε is the random error accounting for noise or variability.

In geopolymer design, LR helps quantify how mix parameters influence the physical and mechanical properties (Matsimbe et al., 2024b). The typical modelling process consists of collecting experimental data for different mix designs (e.g., see Table 4.1), identifying key independent variables influencing the target property, training the regression model to find the best-fitting coefficients that minimize the error between the actual and predicted property values (e.g., using the least squares method, minimizing $\text{error} = \sum_{i=1}^{m}(y_i - \check{y}i)^2$), validating the model using statistical metrics (e.g., coefficient of determination, root mean square error, mean absolute error, mean absolute percentage error), and predicting/optimizing new mix designs without laboratory testing.

For example (see equations 4.2 and 4.3), by inputting a new combination of precursor type and ratio, activator concentration, and curing conditions, the model can estimate the expected compressive strength.

$$f_c = \beta_0 + \beta_1(FA) + \beta_2\left(\frac{\text{Si}}{\text{Al}}\right) + \beta_3\left(\text{NaOH}_{\text{molarity}}\right) + \beta_4(\text{CuringTemp}) \quad (4.2)$$

where f_c is the compressive strength.

After fitting the model with experimental data, suppose the regression equation obtained is:

$$f_c = 6.2 + 0.47(FA) + 1.9\left(\frac{\text{Si}}{\text{Al}}\right) + 0.66\left(\text{NaOH}_{\text{molarity}}\right) + 0.08(\text{CuringTemp}) \quad (4.3)$$

It can be observed that increasing the fly ash content, the NaOH molarity, and the Si/Al ratio influences compressive strength. Furthermore, higher curing temperatures also improve performance but with diminishing returns. The developed model can then guide mix design optimisation by adjusting parameters to achieve the target strength. LR not only predicts outcomes but also provides interpretability, helping engineers understand which variables most strongly influence the geopolymer's behaviour. The regression coefficients quantify the sensitivity of strength to each factor, and the statistical tests (e.g., *t*-test or *p*-values) identify the significant parameters informing experimental focus areas. AI helps optimise the use of industrial waste materials, such as fly ash and phosphogypsum, as precursors for geopolymer production, promoting sustainable construction practices (Matsimbe et al., 2024b). By accurately predicting geopolymer properties, AI can minimize the need for trial-and-error in the laboratory, reducing material waste during the development process.

GPR, also known as Kriging, offers a probabilistic, non-parametric machine learning framework for modelling the complexity and non-linearity of geopolymer design. It predicts target variables and provides uncertainty estimates for each prediction, making it ideal for material science problems where experimental data

TABLE 4.1
Typical input features in geopolymer design

	Binder Composition	Activator Type and Concentration			Curing Condition	Output
Mix	Fly Ash (%)	Na_2SiO_3/ NaOH ratio	Molarity of NaOH	Liquid/ Precursor Ratio	Curing Temp (°C)	Compressive Strength (MPa)
1	100	2.0	10	0.4	80	48.3
2	100	1.5	8	0.3	60	32.5
3	100	1.0	12	0.5	40	27.1

are limited or expensive to obtain. A Gaussian process (GP) defines a distribution over functions $f(x)$, such that any finite collection of function values follows a joint Gaussian distribution:

$$f(x) \sim GP(m(x), k(x, x')) \tag{4.4}$$

where:

- $m(x) = E[f(x)]$ represents the mean function (often assumed to be 0);
- $k(x, x') = E[(f(x) - m(x))(f(x') - m(x'))]$ is the covariance or kernel function that encodes similarity between points x and x'.

For a given dataset $D = \{X, y\}$ with:

- X being the matrix of input parameters (e.g., mix ratio, curing condition);
- y representing the measured property (e.g., strength).

The predictive distribution for a new input x_* is Gaussian with mean and variance represented as:

$$\mu_x = k_*^T (K + \sigma_n^2 I)^{-1} y \tag{4.5}$$

$$\sigma_*^2 = k(x_*, x_*) - k_*^T (K + \sigma_n^2 I)^{-1} k_* \tag{4.6}$$

where:

- K represents the covariance matrix of training data;
- k_* is the covariance vector between the training and testing points;
- σ_n^2 represents the noise variance in measurements.

The kernel function, $k(x, x')$, plays a central role in model development. Common kernels used in geopolymer modelling include:

- radial basis function (RBF), which captures smooth non-linear variations;
- matérn kernel, which models rougher responses;
- polynomial kernel, which models interactions of varying order.

The typical application of GPR in geopolymer design modelling is illustrated in the following steps.

Step 1: Data Preparation

It is required to gather experimental mix design data consisting of the independent variables (e.g., precursor composition, activator concentration, curing conditions) and dependent variables (e.g., strength, workability). All the features should be normalised to ensure they fall within a consistent range.

Step 2: Kernel Selection

It is essential to choose an appropriate kernel function that reflects the smoothness or complexity of the data relationships. For geopolymers, the RBF kernel or Matern kernel is commonly preferred, as they can model the non-linear and continuous relationships between design parameters and resulting strength. A typical example of an RBF kernel is:

$$k(x, x') = \sigma_f^2 \exp\left(-\frac{||x - x'||^2}{2l^2}\right) \quad (4.7)$$

where l controls smoothness, and σ_f^2 controls signal variance.

Step 3: Model Training

Thereafter, it is necessary to train the GPR model by maximizing the log marginal likelihood:

$$\log p(y \mid X) = \frac{1}{2} y^T (K + \sigma_n^2 I)^{-1} y - \frac{1}{2} \log\left|K + \sigma_n^2 I\right| - \frac{n}{2} \log(2\pi) \quad (4.8)$$

This optimization determines the best hyperparameters for the kernel.

Step 4: Prediction

This stage entails predicting the target property for new design combinations. GPR produces the predicted mean (μ_x) and predicted variance (σ_*^2). The uncertainty helps researchers understand which regions of the design space are well explored (low uncertainty) and which need more data (high uncertainty).

Step 5: Model Validation

Thereafter, the developed model is validated using statistical metrics, including the coefficient of determination (R^2), root mean square error (RMSE), and mean absolute percentage error (MAPE). The model's statistical significance is determined using the t-test and p-value. A good GPR model typically achieves an R^2 value > 0.9, p-values < 0.0001, errors within ±20%, and provides reasonable uncertainty intervals around the predicted mean values (Matsimbe et al., 2024a).

Step 6: Optimization

Using GPR, Bayesian optimization is performed to find the optimal combination of mix design parameters that achieves the desired strength or durability target. The probabilistic nature of GPR provides efficient exploration of the design space while minimising the number of laboratory trials.

SVM is a machine learning algorithm based on statistical learning theory used for classification and regression of high-dimensional data and non-linear relationships. It works by finding an optimal boundary (or hyperplane) that best separates data points or fits them within a predefined tolerance, using a subset of the data called support vectors (see Figure 4.1).

For predicting continuous properties, the support vector regression formulation of SVM is used. The goal is to find a function $f(x)$ that deviates from the actual target y_i by at most $\in$, while maintaining the flattest possible function. The mathematical expression is written as:

$$f(x) = w^T \varnothing(x) + b \tag{4.9}$$

where:

- x is the input vector (e.g., mix parameters);
- $\varnothing(x)$ represents the transformation function mapping inputs to a higher-dimensional space;
- w^T is the weight vector to the hyperplane;
- b is the bias term representing the distance of the hyperplane from the origin along the weight vector w.

For a linearly non-separable dataset or outliers, the target is to obtain the hyperplane that maximises the margin between the two classes and ensures the correct classification of all data points. The optimisation problem is written as:

$$\min \frac{1}{2} \| w \|^2 + C \sum_{i=1}^{n} \left(\in_i + \in_i^* \right) \tag{4.10}$$

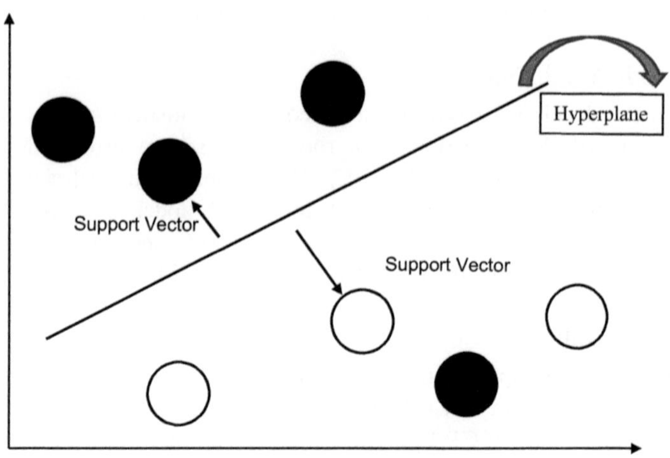

FIGURE 4.1 Optimal hyperplane.

Subject to the constraints:

$$\begin{cases} y_i - \left(w^T \varnothing\left(x_i + b\right) \leq \epsilon + \in_i\right) \\ \left(w^T \varnothing\left(x_i + b\right) - y_i\right) - y_i \leq \epsilon + \in_i^* \\ \in_i, \in_i^* \geq 0 \end{cases} \tag{4.11}$$

where:

- ϵ is the allowable error margin;
- $\in_i, \in_i^*$ are slack variables for deviations beyond ϵ;
- C is the regularisation parameter controlling the trade-off between model flatness and tolerance for error.

To capture the non-linear relationships between mix parameters and material properties, data are mapped into a high-dimensional feature space using kernel functions. The kernel trick allows computations in this space without explicitly performing the transformation. Common kernels used in geopolymer design modelling are illustrated in Table 4.2. For geopolymer data, RBF is the most used because it effectively models smooth, non-linear dependencies. As previously explained, the steps involved in geopolymer design modelling consist of data preparation, model training, model validation, prediction and optimisation, and integration with design software.

The KNN algorithm is one of the simplest, yet powerful, non-parametric machine learning methods used for both regression and classification problems. It serves as an effective data-driven prediction and optimisation tool for geopolymer design modelling. Unlike linear regression or neural networks, KNN does not assume any specific mathematical form between the input and output variables. Instead, it relies on similarity-based reasoning – similar mixes produce similar outcomes. KNN predicts the property of a new mix design by analysing the K closest data points (neighbours) in the training dataset, based on a chosen distance metric. For regression problems (e.g., predicting strength), the expected value for a new data point is calculated as:

$$y'\left(x_{\text{new}}\right) = \frac{1}{K} \sum_{i=1}^{k} y_i \tag{4.12}$$

where y_i are the target values of the K nearest neighbours.

In geopolymer modelling, the Euclidean distance method (equation 4.13) is the most used due to the continuous nature of the mix parameters. Other methods are Minkowski, Manhattan, and Hamming distance. For classification problems (e.g., "pass/fail" for durability), KNN assigns the class most frequently represented among the neighbours.

TABLE 4.2
Common kernels used in geopolymer design modelling

Kernel Type	Function	Usage
Linear	$K\left(x_i, x_j\right) = x_i^T x_j$	Simple, interpretable relationships
Polynomial	$K\left(x_i, x_j\right) = (x_i^T x_j + 1)^d$	Captures polynomial interactions
Radial basis function	$K\left(x_i, x_j\right) = \exp\left(-\frac{\lVert x_i - x_j \rVert^2}{2\sigma^2}\right)$	Maps data into infinite-dimensional space, effective in complex classification problems
Sigmoid	$K\left(x_i, x_j\right) = \tanh(kx_i^T x_j + c$	Mimics neural network activation

$$d = \sqrt{\sum_{i=1}^{k} (x_i - y_i)^2} \tag{4.13}$$

The number of neighbours (K) is a critical hyperparameter. For small K (e.g., 1–3), the model becomes highly sensitive to noise (overfits data). For large K (e.g., >15), the model becomes too smooth (underfits data). The optimal K is typically chosen through cross-validation, trial and error, and/or a common heuristic (i.e., $\sqrt{N}$, where N is the number of training data points). It is recommended to keep N an odd number to avoid ties. Since KNN relies on distance, all variables must be normalised (0–1):

$$x_{\text{norm}} = \frac{x - x_{\text{min}}}{x_{\text{max}} - x_{\text{min}}} \tag{4.14}$$

where: x_{norm} represents the feature after normalization;

x is the original value of the feature;
x_{min} is the minimum value of the feature in the dataset;
x_{max} is the maximum value of the feature in the dataset.

The data normalisation process ensures that variables measured on different scales (e.g., % vs. °C) contribute equally to the distance calculation. For each new mix design, compute the Euclidean distance to all known data points, identify the K closest points, and predict the property as the average (regression) or most common (classification) of these neighbours.

NB algorithm is a probabilistic machine learning method based on Bayes' theorem with the assumption that all input features are independent given the target class. In geopolymer design modelling, NB can be applied to categorise mixes into categories (e.g., low, medium, or high strength), predict curing quality or setting performance, and assess the likelihood of durability (e.g., resistance to sulphate attack). Bayes' theorem forms the foundation of NB models by expressing how to update the probability of a hypothesis class given new evidence data. It is illustrated in the mathematical formula:

$$P(C_i \mid X) = \frac{P(X \mid C_i).P(C_i)}{P(X)} \tag{4.15}$$

where: $P(C_i \mid X)$ refers to the posterior probability – probability of class C_i given features X;

$P(X \mid C_i)$ refers to the Likelihood – probability of observing features X given class C_i;
$P(C_i)$ is the prior probability – initial belief about the class C_i;
$P(X)$ is the evidence – overall probability of observing X.

The algorithm assigns a data point to the class with the highest posterior probability:

$$\check{C} = \text{argmax}_y P(y).\prod_{i=1}^{n} P(C_i \mid X) \tag{4.16}$$

For continuous features, a Gaussian distribution NB classifier is assumed:

$$P(x_i \mid C_i) = \frac{1}{\sqrt{2\sigma_y^2}} \exp\left(-\frac{(x_i - \mu_y)^2}{2\sigma_y^2}\right) \tag{4.17}$$

NB offers a probabilistic, interpretable, and computationally efficient approach to classification and uncertainty modelling in geopolymer design. Although it assumes feature independence, it can deliver high accuracy and meaningful probability estimates even with limited data. When combined with robust models, such as SVM, NB can serve as a baseline model or Bayesian component in hybrid frameworks for predicting robust geopolymer performance.

RF is an ensemble learning algorithm based on the concept of combining multiple DT to produce a more accurate (i.e., low variance) and generalisable predictive model. Each DT provides a prediction, and the forest aggregates these predictions by averaging (for regression) and majority voting (for classification). Compared to the use of a single DT, the RF ensemble approach reduces overfitting and increases robustness and accuracy. Figure 4.2 illustrates the RF algorithm, which comprises bootstrap sampling from the dataset, random feature selection, decision tree training, aggregation, and out-of-bag error estimation (OOB). For a regression task, each DT is averaged, and for a classification task, a majority

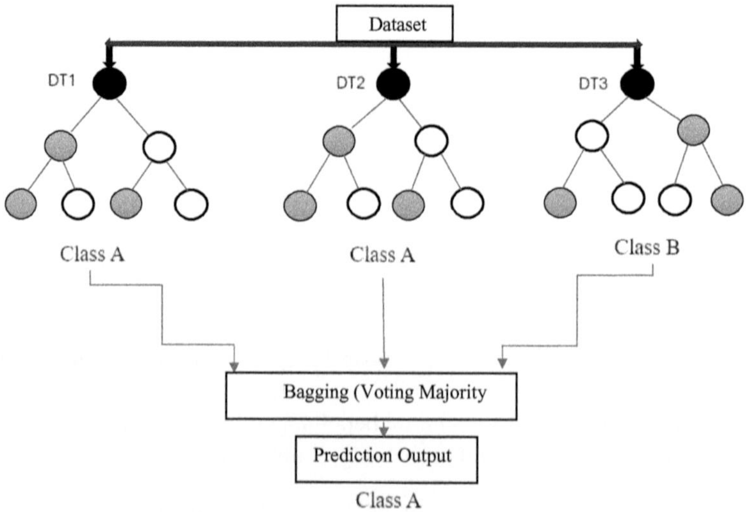

FIGURE 4.2 Schematic illustration of the RF algorithm.

vote yields the predicted class. Finally, the OOB sample is implemented for cross-validation to finalise the prediction work.

Given input features $X = \left[x_1, x_2, \ldots, x_n\right]$ and the target variable y, the RF predictor is mathematically represented as:

$$\hat{f} = \frac{1}{B}\sum_{b=1}^{B} f_b\left(X\right) \tag{4.18}$$

where: B is the number of trees;

$f_b\left(X\right)$ is the prediction from the bth decision tree.

The feature importance is computed as the average decrease in node impurity (e.g., Gini impurity or variance reduction) contributed by each feature across all trees. This computation enables engineers to identify which parameters affect the target property. As part of an integrated machine learning framework, RF enables intelligent optimisation of geopolymer mixes, durability classification, and control of 3D printing parameters, paving the way for data-driven materials engineering (Matsimbe et al., 2024a).

ANN is a computational model inspired by the human brain that excels at learning complex, non-linear relationships from data. Unlike traditional statistical or empirical models that rely on predefined equations, an ANN learns directly from data, discovering hidden correlations between input variables (i.e., mix design parameters) and output variables (i.e., mechanical or durability properties). Figure 4.3 shows a schematic diagram of the ANN.

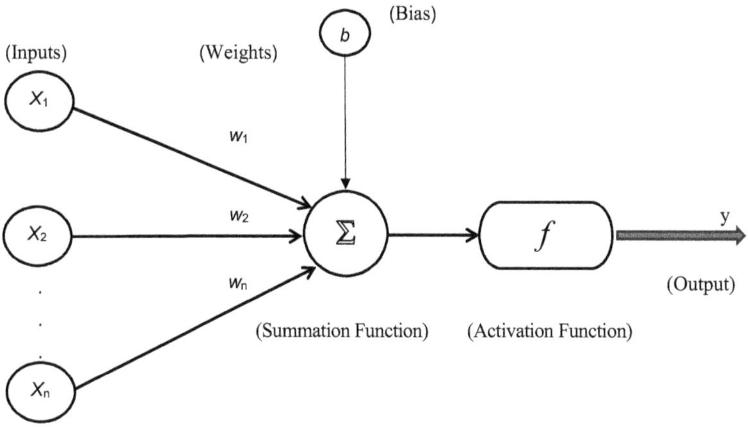

FIGURE 4.3 Schematic diagram of an ANN.

An ANN is composed of interconnected neurons organized in three main layers, consisting of an input layer (receives input data), a hidden layer (performs non-linear transformations using activation functions), and an output layer (produces predicted target values). Each connection between neurons carries a weight that determines the influence of one neuron on another. For a neuron *j* in a hidden layer, the output z_j is given by:

$$z_j = f\left(\sum_{i=1}^{n} w_{ij} x_i + b_j\right) \tag{4.19}$$

where: x_i represents the input features;

$w_{ij} x_i$ represents the connect weights;
b_j is a bias term;
$f(\ )$ is the activation function.

Activation functions play a crucial role in neural networks by introducing non-linearity, enabling the model to capture and learn complex data relationships. Several types of activation functions are commonly employed in ANNs:

- *Sigmoid function:* Produces outputs within the range of 0 to 1 and is typically applied in binary classification problems, such as distinguishing between two categories (e.g., identifying whether an image contains masonry or not).
- *ReLU (rectified linear unit):* Widely used in hidden layers, this function outputs the input value if it is positive and zero otherwise. It is particularly

effective in addressing the vanishing gradient issue, thereby improving learning efficiency.

- *Tanh (hyperbolic tangent):* Similar in nature to the sigmoid function but generates outputs between −1 and 1. It is preferred in hidden layers where both positive and negative activation values are beneficial.
- *Softmax function:* Converts a set of raw output scores into normalised probability values, making it ideal for multi-class classification tasks in the output layer.
- *Leaky ReLU:* A modified version of ReLU that allows a small, non-zero gradient for negative input values, helping to prevent the occurrence of "dead neurons" during the training process.

Overall, activation functions determine whether a neuron becomes active or not, allowing neural networks to detect intricate patterns within data and make reliable predictions. The training phase adjusts weights and biases to minimise prediction errors using a loss function. For regression problems in geopolymer modelling, the mean squared error is common:

$$L = \frac{1}{n}\sum_{i=1}^{n}(y_i - \widehat{y}_i)^2 \tag{4.20}$$

Optimisation is performed using the backpropagation algorithm, where gradients of the loss function are computed based on each weight. In tandem, the weights are updated via gradient descent or the Adam optimiser:

$$w_{\text{new}} = w_{\text{old}} - \alpha\frac{\delta L}{\delta w} \tag{4.21}$$

4.4.2 Addressing the Heterogeneity of Geopolymers

Geopolymer mix design involves complex interactions among multiple variables – aluminosilicate precursor composition, activator concentration and composition, curing conditions, and additives – which influence reactions chemically non-linear and coupled, e.g., dissolution kinetics, geopolymerisation rate, and microstructural development. Traditional methods, such as Taguchi and RSM, cannot accurately capture these non-linear dependencies. Machine learning models overcome this limitation by utilising non-linear activation functions, kernel transformations, ensemble averaging, and hierarchical feature learning, enabling them to approximate any complex function given sufficient data. By leveraging AI and ML, researchers and engineers can accelerate the discovery, development, and adoption of different geopolymer materials, contributing to more sustainable and efficient construction practices. Rather than replacing experiments, AI/ML can be used as a complementary tool for prescreening compositions before laboratory trials, sensitivity analysis to identify the most influential parameters, and reducing experimental load by focusing on promising formulations.

While ML models have demonstrated remarkable accuracy in predicting geopolymer properties and addressing geopolymer heterogeneity, their implementation has largely been confined to research environments through using software such as Python, MATLAB, or R. These ML research environments are supported by ML frameworks/libraries, e.g., Scikit-learn, PyTorch, TensorFlow, RapidMiner, Google Colab, and Ludwig. For widespread industrial adoption, ML capabilities must be integrated into commercial design and simulation software platforms commonly used by civil engineers and material scientists, such as ABAQUS, COMSOL, ANSYS, RocScience, ConcreteLab, Midas, or OriginLab. Such integration enables researchers and engineering practitioners to simulate real-time material performance based on AI-driven predictions, optimise mix designs directly within engineering workflows, conduct parametric sensitivity analysis, and automate mix proportioning. It also allows for coupling ML predictions with finite element analysis for structural simulations. For example, a geopolymer mix database stored in Excel or SQL can be linked via a Python API to COMSOL, allowing the ML model to update material parameters based on user-defined mix inputs. In COMSOL, the ML models can be imported via *LiveLink* for MATLAB/Python, allowing predictive models to define material constitutive laws. In ABAQUS, ML-based material models can be integrated using the *UMAT* or *VUMAT* function. In ANSYS, the *DesignXplorer* environment supports AI-based design optimisation and surrogate modelling.

An integrated geopolymer design modelling framework should comprise data acquisition, ML model training/testing, deploying/embedding the trained model into commercial software, and simulating/conducting virtual experiments, structural analyses, or parametric sweeps using AI-predicted material parameters. This hybrid modelling framework bridges the gap between laboratory data science and practical engineering applications. It significantly reduces the experimental testing requirements while improving predictive accuracy and design flexibility. Table 4.3 summarises extended functionalities enabled by integrating ML into commercial geopolymer design tools.

Future AI/ML prospects should consider integrating AI with real-time data from sensors embedded in geopolymer structures to create digital twins that can help monitor and predict performance throughout the lifecycle. Furthermore, the rise of autonomous labs, where AI controls robotics for mixing, curing, and testing, is expected to revolutionise geopolymer research and development by accelerating iteration and discovery cycles. Complementary research should focus on developing open-source ML-FEA frameworks and collaborative databases to enable reproducible, transparent geopolymer design modelling. To overcome data-sharing barriers, federated learning enables multiple institutions to train shared models without transferring proprietary data, thereby ensuring collaboration and maintaining data privacy. However, combining domain expertise, high-quality data, and ethical considerations is essential to harness these technologies.

TABLE 4.3
Extended functionalities enabled by ML integration

Extended Feature	Description	Enabled by ML Integration
Adaptive material libraries	Databases that automatically update with new experimental data	Continuous learning models
Automated mix optimisation	Real-time optimisation of mix proportions	Genetic algorithms, Bayesian optimisation
Predictive durability modelling	Forecasting long-term performance under chemical or thermal attack	Recurrent neural network
Microstructure-property mapping	Linking SEM/XRD data with macro-properties	Convolutional neural network
Real-time digital twin models	Simulation of geopolymer curing and strength development in real time	Reinforcement learning

4.5 STANDARDS, CODES, AND REGULATIONS

Geopolymer chemistry, variability of precursor materials, and differing production/curing techniques make direct adoption of OPC-centric standards difficult. The absence of globally accepted international standards for structural geopolymer binders remains a significant barrier to their widespread adoption (Matsimbe et al., 2024b). Standards and codes offer numerous advantages for industrial acceptance, as they ensure safety and reliability, foster market confidence, promote regulatory compliance, and facilitate interoperability. Standardised tests and acceptance criteria ensure that materials meet minimum performance and durability for structural use. This has an impact on contractors and regulators who become more willing to adopt materials that fit within established codes. Furthermore, this assures that environmental safety and product labelling requirements are enforced. Standard test methods allow designers and specifiers to compare products and make performance-based choices. As of 2025, no single dedicated ISO or EN standard fully covers geopolymer binders for structural applications. This lack of a bespoke international standard has hindered the harmonisation of adoption. EN 206 and EN 197 were historically written for ordinary Portland cement, technically excluding geopolymer binders unless under special provisions. ASTM has hosted symposia and technical work on geopolymer/alkali-activated binders, which have culminated in the development of the *Standard Specification for Alkali-Activated Cementitious Materials* (ASTM C1948/C1948-24 [ASTM C1948/C1948-24, 2024]) and *Standard Test Method for Compressive Strength of Alkali-Activated Cementitious* (ASTM C1928/C1928M-25 [ASTM C1928/C1928M-25, 2025]). RILEM technical committees have been a major vehicle for developing recommendations for testing and performance assessment of alkali-activated materials and/or geopolymers. RILEM's working groups publish technical recommendations that are often used as the scientific basis for later standardisation by national and international bodies.

Alkali activators are caustic and require clear occupational safety rules: PPE, spill response plans, handling and storage specifications, training for mixing personnel, and emergency procedures. Many jurisdictions already regulate the handling of caustic chemicals – specifications should cross-reference national safety legislation (e.g., COSHH regulations). Because geopolymer precursors are industrial by-products, regulatory assessment of material provenance, potential contaminants, and lifecycle claims is required. Project specifications that accept waste-derived feedstocks require documentation of traceability and chemical characterisation. To promote the advancement of standards, codes, and regulations, it would be ideal to adopt a performance-based approach on projects, adopt material traceability and certification for precursors/activators, document production and curing procedures, reinforce environmental and safety plans, promote project-specific QA/QC, conduct pilot testing, and engage regulatory acceptance bodies and building control officers during specification development to identify additional local testing or certification requirements. Various stakeholders, comprising researchers, engineers, standards committees/regulators, and policymakers, must continue coordinated efforts to close gaps in long-term durability data, harmonise testing methods, and reconcile environmental/waste regulation issues.

4.6 3D PRINTING WITH GEOPOLYMERS

The construction industry is currently undergoing a technological revolution driven by digital fabrication, automation, and sustainability imperatives. Among the most disruptive technologies is 3D printing, also known as additive manufacturing (Sanjayan et al., 2019). In civil engineering and materials science, 3D printing of cementitious materials has received significant attention because it enables automated construction of complex geometries without formwork, minimizes material waste, shortens construction timelines, and reduces labour costs. Geopolymers have emerged as a sustainable alternative to Portland cement in 3D printing applications. Integrating geopolymer technology with 3D printing has the potential to transform sustainable construction, addressing both environmental and technological challenges. However, tailoring the rheological and setting behaviour is critical to achieve pumpability, extrudability, buildability, mechanical strength, interlayer bonding, and durability. The most common method for 3D printing (see Figure 4.4) using geopolymers involves extrusion of the fresh mix through a nozzle, layer by layer.

FIGURE 4.4 3D Printing in progress at the University of Johannesburg.

The key 3D printing process parameters include nozzle diameter, layer height, printing speed, and extrusion pressure (Sanjayan et al., 2019). Gantry systems are well-suited for large-scale elements, e.g., walls and columns. Robotic arms allow more freedom in printing complex geometries. For extrusion-based additive manufacturing, geopolymers exhibit several favourable attributes, including rapid setting and early strength gain (ensuring buildability), low shrinkage and creep (reducing deformation during printing), compatibility with industrial waste materials (aligning with sustainability goals), and fire and chemical resistance (ideal for specialized applications). Three-D-printed geopolymers have various applications in both structural and non-structural elements. They provide rapid deployment of housing units using locally available raw materials, making them highly suitable for disaster relief and affordable housing in developing countries. For example, in April 2023, three companies – Geopolymer International, Renca, and StrongPrint3D – pioneered the world's first 3D-printed geopolymer house, having dimensions 10 m × 3 m × 2.6 m.

3D printable geopolymers should exhibit fresh-state properties that enable the mix to flow easily through hoses and nozzles without segregation/blockage. The printed layers should retain their shape and form without excessive deformation or collapse under self-weight. The material should remain printable for a sufficient period to support continuous operations, but it must be able to set in a reasonable time once deposited. Furthermore, the 3D printable geopolymers should exhibit adequate compressive and flexural strength that meet structural requirements. There must be strong adhesion between successive layers to prevent delamination. Satisfactory fresh- and hardened-state properties depend on the selection of appropriate precursors, activators, and/or additives. Compared to slag-based geopolymers that offer rapid setting and high early strength, low-calcium fly ash–based geopolymers provide good workability but slower strength gain (Matsimbe et al., 2022). Hybrid mixes (fly ash + slag or fly ash + metakaolin) provide a balance of printability and performance. Adjusting the SiO_2/Na_2O ratio in the activator modifies viscosity, setting time, and final strength. The presence of fibres, such as polypropylene, in the geopolymer mixture has the potential to enhance both the tensile strength and interlayer bonding.

PROBLEMS

1. Explain the difference between the trial-and-error and Taguchi approaches in geopolymer design.
2. Why might the trial-and-error method still be used in small-scale geopolymer research?
3. Given three variables: NaOH molarity, Si/Al ratio, and curing temperature. Outline how you would design a Taguchi L9 array and the impact of the S/N ratio where higher strength is desirable.
4. Describe how you can develop a regression-based model for geopolymer compressive strength using Python.

5. Explain how you can apply a genetic algorithm to optimise Si/Al and Na_2O/Al_2O_3 ratios for maximum durability.
6. Based on your experience in geopolymer modelling, propose an experimental plan integrating Taguchi and RSM to minimise the drying shrinkage of geopolymers.
7. With the aid of well labelled sketches and equations, evaluate how AI and ML are transforming geopolymer mix design and performance prediction.
8. A geopolymer mix used for 3D printing collapses after extrusion. Provide explanations for the following:
   - Identify two probable causes of poor buildability.
   - Suggest two practical adjustments to the mix design to improve shape stability.
9. You are tasked with measuring the yield stress (τ_0) and plastic viscosity (μ_p) of a geopolymer printing paste using a rotational rheometer.
   - Explain how these parameters influence pumpability and layer stability.
   - Discuss how temperature and activator concentration affect these parameters.
10. A 3D printer extrudes a geopolymer paste at a rate of 10 kg/min. The open time of the mix is approximately 25 minutes before it begins to harden.
    - How much material (in kilograms) can theoretically be printed before the mix becomes unworkable?
    - If the printing process is delayed by 15 minutes, discuss what physical effects may occur in the mix and how they could be mitigated.
11. Describe three factors that influence the bond strength between successive layers in 3D printed geopolymer elements. Explain how these can be improved during the printing process.

REFERENCES

ASTM C1928/C1928M-25. (2025). *Standard Test Method for Compressive Strength of Alkali-Activated Cementitious.* ASTM International, West Conshohocken, PA, USA.

ASTM C1948/C1948-24. (2024). *Standard Specification for Alkali-Activated Cementitious Materials.* ASTM International, West Conshohocken, PA, USA.

Matsimbe, J., Dinka, M., Olukanni, D., and Musonda, I. (2022). Geopolymer: A systematic review of methodologies. *Materials*, vol. 15, no. 19.

Matsimbe, J., Dinka, M., Olukanni, D., and Musonda, I. (2024a). Fundamental machine learning algorithms and statistical models applied in strength prediction of geopolymers: A systematic review. *Discover Applied Sciences*, vol. 6, no. 538.

Matsimbe, J., Dinka, M., Olukanni, D., and Musonda, I. (2024b). Performance evaluation and mix design of ambient-cured fly ash-phosphogypsum blended geopolymer paste and mortar. *Results in Engineering*, vol. 24, no. 103280.

Sanjayan, J.G., Nazari, A., and Nematollahi, B. (2019). *3D Concrete Printing Technology: Construction and Building Applications.* Butterworth-Heinemann, Kidlington, UK.

5 Geopolymer Practical Applications

5.1 SPECIFIC CASE STUDIES

The global case studies detailed in the subsequent sections show that geopolymer technology is fit for mainstream adoption across a wide range of industrial applications. These positive, successful projects demonstrate that, with the right precursor materials, mix design, curing protocols, training, safety policies, and standardisation, geopolymers can complement traditional Portland cement in terms of performance, durability, and sustainability.

Case Study 1: Brisbane West Wellcamp Airport (Toowoomba, Queensland, Australia)

The Wellcamp Airport project, undertaken in 2014, utilized geopolymer concrete for large areas of heavy-duty pavements (i.e., apron, turning node, taxiways), plus a range of civil works (i.e., kerbs, culverts, precast panels, footings). Published figures report roughly 30,000–40,000 m^3 (70,000–100,000 tonnes) of geopolymer concrete placed (Matsimbe et al., 2022; Glasby et al., 2015; Geopolymer Institute, 2014). Wagners' objective was to demonstrate a low-carbon alternative to ordinary Portland cement concrete for major infrastructure development. The project is one of the first and largest commercial applications of modern geopolymer concrete, delivering 6600 tonnes of CO_2 savings compared to the OPC equivalent for the works using earth-friendly concrete (EFC).

EFC is a cement-free binder comprised primarily of industrial by-products, i.e., pulverised fly ash (class F) blended with ground granulated blast furnace slag (GGBFS). The aluminosilicate precursors replace Portland cement and provide the geopolymeric binder when activated. Commercial geopolymer mixes typically use an alkaline solution (sodium silicate + sodium hydroxide) or equivalent activator system to dissolve aluminosilicates and form the geopolymer gel. Wagners' published project reports describe the use of engineered activator systems adapted for field batching (Glasby et al., 2015).

Airport pavements sustain heavy loads and frequent aircraft movements, and geopolymer technology demonstrated robust mechanical performance, low

DOI: 10.1201/9781003787389-5

permeability to protect embedded steel, and reliable in-service performance in a subtropical environment. The pavements targeted structural grades comparable to or exceeding conventional airport concrete; the typical compressive strength is approximately 40 MPa, and the flexural strength reaches 4.8 MPa at 28 days. Laboratory and field testing also showed low permeability and good resistance to chemical attack compared to OPC mixes. The successful implementation of this project validated the application of geopolymer at an industrial scale.

Case Study 2: University of Queensland Global Change Institute Building (Brisbane, Australia)

The Global Change Institute (GCI) building at the University of Queensland (UQ), completed in 2012, is a landmark project showcasing sustainable construction and innovative materials engineering. A 6-Star Green Star-rated research facility, the building was conceived as a living laboratory for climate-responsive architecture. Its goal was to demonstrate technologies that reduce embodied energy, operational energy, and environmental footprints. One of the building's signature innovations was the use of geopolymer precast concrete elements. At the time, the use of geopolymer concrete in large-scale structural components was still emerging globally. UQ sought to demonstrate that geopolymer-based materials could replace traditional Portland cement–based precast elements, especially in climates where low-carbon, durable materials were needed. The GCI building thus became one of the first four-storey commercial buildings in the world to incorporate structural geopolymer precast concrete as a major component of its façade and floor systems.

The GCI project team, comprising HASSELL architects, Bligh Tanner Consultants, and Wagners, adopted fly ash–based geopolymer concrete as the binder system to achieve material performance and sustainability targets. The building integrated geopolymer concrete in three suspended floors and 33 beams of 10.5 m by 2.4 m. Each element was engineered to meet structural load requirements, resist weathering, and maintain aesthetic quality. The geopolymer precast concrete was designed to achieve compressive strengths of approximately 40 MPa, maintain appropriate modulus of elasticity and thermal properties, display adequate flexural performance for façade loading, and provide long-term durability with reduced cracking (Glasby et al., 2015). The structural design was based on Australian Standards, with mechanical test data from Wagners confirming that the geopolymer mix design met specifications for structural concrete applications.

Case Study 3: Transnet City Deep Container Terminal in South Africa

Transnet City Deep is South Africa's largest inland container terminal and a critical freight gateway for Gauteng. Between 2011 and 2015, Transnet implemented an extensive upgrade and resurfacing program to expand capacity and modernise the terminal. The upgrade required resurfacing 144,000 m^2 of heavy-duty concrete pavement and the recycling of a large volume of existing concrete pavement. The project had explicit sustainability targets of reducing embodied carbon, diverting construction waste from landfill, and creating a strong incentive to trial low-cement alternatives.

Murray & Roberts Infrastructure (the main contractor) and technical partners such as ARC Innovations developed and supplied alternative concrete technologies. The project incorporated long-term test sections of geopolymer concrete, the first commercial cast of 0% ordinary Portland cement concrete in South Africa, on 13 September 2013, to validate performance under real terminal traffic and environmental exposure. These activities were part of a broader industry push, supported by Transnet, to trial low-carbon concretes for heavy structural pavements. Large trial sections utilized a full cement replacement geopolymer blend (60% PFA + 40% GGBFS activated by an alkaline activator). These sections were designed to justify the 100% cement replacement combined with local aggregates and recycled materials. The reported 28-day compressive strengths for the geopolymer concrete were around 51 MPa (Matsimbe et al., 2022). Given the high carbon intensity of ordinary Portland cement, substituting PFA/GGBFS offered a significant carbon saving per cubic metre of concrete.

Case Study 4: NTPC/NETRA and CSIR: India's Pioneering Geopolymer Concrete Roads

India's earliest publicly documented, large-scale field use of cement-free geopolymer concrete for transportation infrastructure was led by National Thermal Power Corporation (NTPC) through its research arm NETRA in partnership with CSIR research institutes. Beginning with trial patches at CSIR-CBRI Roorkee and NTPC-Dadri (2016–2018) and followed by longer pilot stretches (Ramagundam and later NTPC plant roads), NTPC's program delivered M40 geopolymer pavement trials, attracted the attention of Indian Roads Congress (IRC), and culminated in IRC accreditation for NETRA's geopolymer road technology. The initiative pioneered the practical protocols to be followed for precursor material selection, mix design, on-site placement, and quality assurance for geopolymer pavements in India.

Case Study 5: Wagners/Keltbray Group/Capital Concrete Piling Project in the UK

In January 2020, Wagners supplied and placed an industrial-scale EFC geopolymer concrete pile in London: a 900 mm diameter, 25 m deep anchor pile installed for Keltbray Group and Capital Concrete at the Landsec Nova East project. The anchor pile had standard geotechnical and structural performance requirements for an urban, deep-foundation element, i.e., axial capacity, lateral stiffness, and durability, suitable for the expected lifetime of the structure. The EFC mix had to match standard concrete performance for workability, pumpability, early and long-term strength, and durability while delivering lower embodied CO_2. The client and contractors required that the pile be compatible with conventional piling practice and meet the acceptance testing regime typical of piling contracts. EFC was engineered to meet equivalent compressive strength classes and to behave similarly to Portland concrete in structural calculations. The Nova East pile met the acceptance test criteria applied to conventional piles.

5.2 IMPACT ON SUSTAINABLE DEVELOPMENT GOALS

The global shift toward sustainability has spurred interest in geopolymers as they reduce carbon emissions and promote circular economy principles. This section situates geopolymer development within the broader framework of the United Nations SDGs and demonstrates how it advances global efforts toward sustainability. Figure 5.1 illustrates a flowchart that demonstrates how the key UN SDGs (i.e., 9, 11, 12, and 13) are advanced through the adoption of sustainable geopolymers.

In advancing SDG 9 (industry, innovation, and infrastructure), geopolymers support the shift to sustainable industrial processes by valorising industrial waste

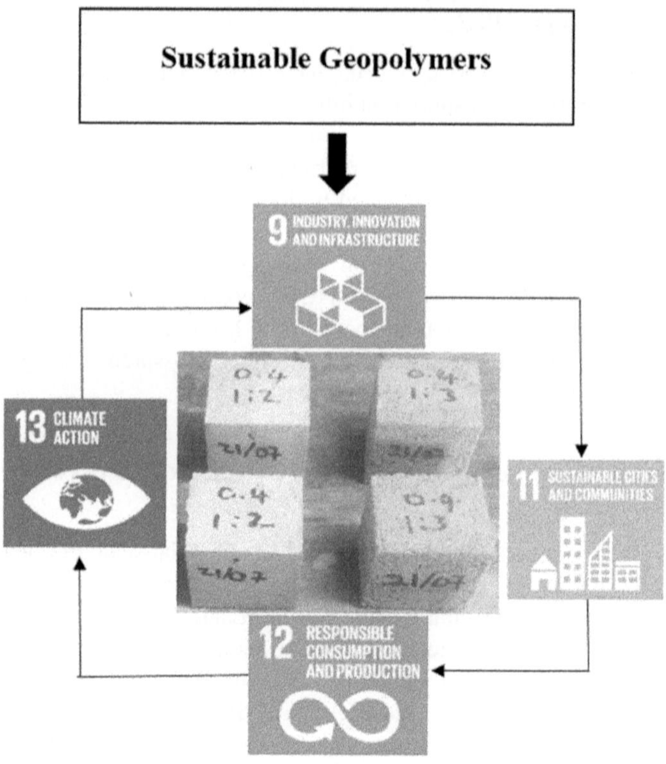

FIGURE 5.1 Flowchart showcasing the key SDGs advanced by sustainable geopolymers.

materials such as fly ash, slag, phosphogypsum, and mine tailings. Unlike OPC, which relies on the extraction of virgin materials for its feedstock (see Figure 5.2), the production of geopolymers encourages industrial symbiosis by transforming waste streams into feedstocks and promoting local resource utilisation, thereby reducing dependence on clinker for OPC. In addition, geopolymers stimulate innovation in material science and enhance infrastructure resilience by exhibiting excellent resistance to fire, chemicals, and sulphate attack (Provis and van Deventer, 2009; Matsimbe et al., 2024), making them ideal for critical infrastructure in aggressive environments.

In advancing SDG 11 (sustainable cities and communities), geopolymers provide green building materials for urban sustainability and contribute to sustainable construction by reducing embodied carbon in buildings, improving indoor air quality due to low emissions, and enhancing thermal and acoustic performance. Furthermore, geopolymers can be utilised in low-cost housing initiatives, particularly in disaster-prone or resource-constrained regions. Their durability (Matsimbe et al., 2024) and minimal maintenance make them suitable for informal settlements and post-disaster reconstruction.

In advancing SDG 12 (responsible consumption and production), geopolymer technology exemplifies responsible production by converting waste into value-added products, reducing the extraction of virgin materials, and lowering water and energy consumption. Compared to ordinary Portland cement, geopolymers demonstrate lower lifecycle carbon emissions, reduced water usage during manufacture, and higher durability (Provis and van Deventer, 2009), thereby

FIGURE 5.2 Limestone drilling in progress for OPC downstream feedstock.

extending the service life of structures and reducing the frequency of repairs. The development of geopolymer-specific standards and Environmental Product Declarations (EPDs) will further promote transparency and accountability in sustainable construction.

In advancing SDG 13 (climate action), geopolymers reduce greenhouse gas emissions through clinker-free production processes, lower calcination temperatures, and avoidance of fossil-fuel-intensive sintering. Infrastructure built with geopolymers is more resilient to climate-induced stresses such as flooding, chemical corrosion, wildfires, and soil salinity (Matsimbe et al., 2024). The widespread adoption of geopolymer technologies supports countries' Nationally Determined Contributions (NDCs) under the Paris Agreement by mitigating emissions in the construction sector, advancing low-carbon industrial strategies, and promoting the transfer of clean technology and capacity building.

In summary, geopolymer development embodies the principles of sustainable development through innovation, waste reduction, and the development of climate-resilient infrastructure. By aligning with SDGs 9, 11, 12, and 13, geopolymers provide a pathway for countries to reduce their ecological footprints while fostering inclusive industrial and urban growth. The time is ripe to scale up geopolymer adoption globally, backed by science, policy, and public awareness.

5.3 IMPACT ON CIRCULAR ECONOMY

The traditional linear economic model of "take, make, dispose" has proven unsustainable in the face of climate change, resource depletion, and waste accumulation (Matsimbe et al., 2022; Provis and van Deventer, 2009). The circular economy (CE), by contrast, is an economic system that eliminates waste and promotes the continual use of resources through reuse, refurbishment, recycling, and sustainable design. Geopolymer development, rooted in the reuse of industrial waste, is a prime example of CE principles in action. As an alternative to ordinary Portland cement, geopolymers offer a low-carbon, resource-efficient solution that supports closed-loop cycles within the construction industry. The core principles defining CE comprise designing out waste and pollution, keeping products and materials in use, and regenerating natural systems. In the context of construction materials, these principles translate into reducing reliance on virgin raw materials; utilising secondary materials; designing for durability, disassembly, and reuse; and minimising environmental impacts throughout their lifecycle.

Geopolymers are typically synthesized using aluminosilicate-rich industrial waste materials, e.g., fly ash, slag, phosphogypsum, red mud, rice husk ash, palm oil fuel ash, sugarcane bagasse ash, mine tailings, and demolition waste. By converting these waste materials into value-added products, geopolymer development directly supports CE objectives by reducing landfill demand, conserving virgin resources, and reducing the carbon intensity of cementitious materials. Geopolymer production consumes less water and lower energy than Portland cement, especially since it bypasses the high-temperature calcination process. This efficient use of materials and energy aligns with the CE focus on reducing input flows while maximising output value. Due to their high chemical resistance, fire resistance, and mechanical durability (Hardjito and Rangan, 2005; Provis and van Deventer, 2009), geopolymers offer longer service life for infrastructure components, lower maintenance and replacement rates, improved performance in aggressive environments (Matsimbe et al., 2024), and enhanced material utility per lifecycle, reducing the need for premature demolition and waste generation.

Geopolymers are used in precast blocks, bricks, pavers, concrete panels, and 3D-printed structural elements, which support CE models by enabling modular design, deconstruction, and reuse. Old or degraded infrastructure can be repaired using geopolymer mortars or coatings, thereby extending the life of existing assets, reducing embodied emissions, and preventing the need for full replacement. Geopolymer production fosters industrial symbiosis, where the waste output of one industry serves as the raw input for another (Hardjito and Rangan, 2005; Provis and van Deventer, 2009). For example, a coal-fired power plant produces fly ash useful in a nearby geopolymer brick factory; a steel mill supplies slag to a precast concrete producer; and agricultural residues are sourced to produce lightweight geopolymer composites.

PROBLEMS

1. Propose a geopolymer mixture combination suitable for producing paving blocks in a tropical climate using locally available industrial waste materials.
2. Evaluate the potential barriers to large-scale commercialisation of geopolymers in the world. Discuss solutions based on lessons from global case studies.
3. Using the Toowoomba Wellcamp Airport case, calculate the estimated CO_2 reduction per cubic meter of geopolymer concrete if OPC concrete emits 306 kg CO_2/m^3.

REFERENCES

Geopolymer Institute. (2014). *70,000 Tonnes Geopolymer Concrete for Airport*. Institut Géopolymère, Saint-Quentin, France.

Glasby, T., Day, J., Genrich, R., and Kemp, M. (2015). Commercial scale geopolymer concrete construction. In *Proceedings of the Saudi International Building and Construction Technology Conference*, Riyadh, Saudi Arabia, 11–12 May.

Hardjito, D., and Rangan, B.V. (2005). *Development and Properties of Low-Calcium Fly Ash-Based Geopolymer Concrete*. Curtin University of Technology, Australia.

Matsimbe, J., Dinka, M., Olukanni, D., and Musonda, I. (2022). Geopolymer: A systematic review of methodologies. *Materials*, vol. 15, no. 19.

Matsimbe, J., Dinka, M., Olukanni, D., and Musonda, I. (2024). Durability properties of ambient-cured fly ash-phosphogypsum blended geopolymer mortar in terms of water absorption, porosity, and sulfate resistance. *Discover Sustainability*, vol. 5, no. 330.

Provis, J., and van Deventer, J. (2009). *Geopolymers: Structures, Processing, Properties and Industrial Applications*. Woodhead Publishing Ltd, Cambridge, UK.

6 Geopolymer Lifecycle Assessment

6.1 GLOBAL WARMING POTENTIAL

The construction industry is a major contributor to global greenhouse gas emissions, responsible for nearly 40% of global CO_2 emissions, primarily through the production and use of ordinary Portland cement (Matsimbe et al., 2022). As the world seeks sustainable alternatives to mitigate climate change, geopolymers have emerged as a promising substitute due to their capacity to reduce the impact of construction materials on the environment. According to ISO 14040/14044 (ISO 14040, 2006) standards, the LCA procedure comprises goal and scope definition, lifecycle inventory, lifecycle impact assessment, and interpretation (Hauschild et al., 2018). Global Warming Potential (GWP) is a measure used to compare the ability of different greenhouse gases to trap heat in the atmosphere. It is expressed relative to CO_2, which has a GWP of 1. In the context of construction materials, GWP is commonly assessed via LCA and reported in terms of CO_2-equivalents (CO_2-eq) per unit mass or volume of material. To accurately compare materials, LCA must define system boundaries as cradle-to-gate (raw material extraction to product manufacture), cradle-to-grave (extending to product use and disposal), and cradle-to-cradle (recycling or reuse at the end of life).

OPC production is highly energy intensive and involves two major sources of CO_2 emissions: the calcination process, where limestone ($CaCO_3$) is heated to produce lime (CaO), releasing CO_2; and the combustion of fossil fuels used to heat the kiln during clinker production (see Figure 6.1). The average CO_2 emission from OPC production is estimated to be greater than 0.6 kg CO_2/kg cement (Provis and van Deventer, 2009; McLellan et al., 2011; Turner and Collins, 2013). Given the scale of global OPC consumption, this has a significant contribution to anthropogenic CO_2 emissions. Comparatively, geopolymers are produced through the alkaline activation of aluminosilicate-rich precursor materials. Because geopolymers rely on industrial waste materials and eliminate the need for calcination, their GWP is substantially lower than that of OPC.

While geopolymers avoid clinker production, GHG emissions arise from the mining and transportation of raw materials, the manufacture and use

DOI: 10.1201/9781003787389-6

FIGURE 6.1 Clinker production infrastructure.

of alkaline activators, and curing processes, especially when heat curing is applied. Nonetheless, GWP values for geopolymers range from 0.15 to 0.50 kg CO_2/kg geopolymer binder, depending on the materials and process route, which is up to 70% lower than that of OPC (Provis and van Deventer, 2009; Habert and Ouellet-Plamondon, 2016; Matsimbe et al., 2022). The key factors affecting the GWP of geopolymers comprise the type of precursor material, type and concentration of alkaline activators, the transportation distance, and curing method. Fly ash and slag are industrial by-products with low embodied emissions compared to metakaolin, which has a higher GWP due to energy use during kaolin clay calcination.

6.2 CUMULATIVE ENERGY DEMAND

Cumulative Energy Demand (CED) is a lifecycle-based metric that measures the total direct and indirect energy input required to produce a material or system, expressed in megajoules (MJ) per functional unit (e.g., per kilogram of binder, per cubic meter of concrete). CED constitutes the direct energy consumed during raw material processing, transportation, mixing, and curing, and the indirect energy embedded in the production of input materials such as NaOH, Na_2SiO_3, fly ash, or slag. Furthermore, CED can be categorized into non-renewable fossil energy (coal, oil, and natural gas), non-renewable nuclear energy, and renewable energy resources (solar, hydro, wind, and biomass). The construction industry accounts for a significant share of global energy consumption, largely due to the production of cement and other conventional binders. OPC is highly energy intensive, with production involving calcination at around 1400°C and extensive use of fossil fuels (Turner and Collins, 2013). In the pursuit of sustainable alternatives, geopolymers have emerged as promising materials with a lower environmental burden. One of the most critical aspects of assessing the environmental performance of geopolymers is CED. By evaluating the CED of geopolymers, researchers and policymakers can compare their energy efficiency with that of conventional cementitious materials and identify opportunities for improvement in formulation, production, and application.

CED for geopolymers is assessed in terms of aluminosilicate precursors, alkaline activators, and curing processes. FA, commonly used as a precursor for geopolymers, is a by-product of coal combustion, and its production energy is typically allocated as negligible. However, drying, grinding, and transport may add minor energy inputs. GGBFS is a by-product of steelmaking, with energy allocation depending on system boundary assumptions. The energy demand for GGBFS emanates from granulation and grinding. In contrast, metakaolin requires the calcination of kaolinite clay at 600–800°C, resulting in significant direct energy consumption compared to fly ash or slag. NaOH, a commonly used alkaline activator, is highly energy intensive, with CED up to 3.5 MJ/kg due to electrolysis of brine. Similarly, Na_2SiO_3 contributes significantly to the CED of geopolymer systems, with values greater than 5 MJ/kg and production temperatures exceeding 1300°C, resulting from the melting of silica sand and sodium carbonate.

Ambient curing offers low energy demand in line with SDG 13, compared to heat curing, which significantly increases the CED. Overall, compared to OPC, geopolymers have reduced CO_2 emissions and comparable or higher CED than OPC, depending on the mix design. For example, OPC binder has a CED of approximately 4–6 MJ/kg of cement, while geopolymers have a CED ranging from 5–12 MJ/kg. Although the CED of geopolymers can be high due to the production of alkaline activators, their GWP is generally much lower than that of OPC, indicating a trade-off between energy intensity and carbon footprint (Matsimbe et al., 2022; Turner and Collins, 2013). Sodium silicate activator usually dominates CED; therefore, reducing its content lowers the total energy demand. With ongoing advancements in renewable energy, waste valorisation, and material optimisation, geopolymers hold strong potential as a sustainable alternative binder for the future of construction.

6.3 ACIDIFICATION POTENTIAL

Sustainable development in the construction sector requires comprehensive environmental assessments of building materials beyond greenhouse gas emissions and energy use. One critical environmental impact category is Acidification Potential (AP), which measures the potential of emissions to contribute to the acidification of soil and water systems. Acidification occurs when nitrogen oxides (NO_x), sulphur dioxide (SO_2), ammonia (NH_3), and other acidifying gases are released into the atmosphere. These pollutants react with water and oxygen to form acids such as sulphuric acid (H_2SO_4) and nitric acid (HNO_3). When deposited onto land or water, these acids can cause ecological damage, including soil degradation, reduced agricultural productivity, forest decline, and loss of aquatic biodiversity.

In ordinary Portland cement production, AP emanates from fossil fuel combustion during clinker production (emits SO_2 and NO_x) and electricity usage in grinding and/or material handling. In contrast, geopolymers, synthesised from industrial by-products such as FA, GGBFS, or metakaolin, offer potential reductions in acidification impacts. However, the production of alkaline activators remains energy-intensive and associated with emissions that influence AP, with contribution values ranging from 0.01 to 0.0051 kg SO_2-eq/kg binder (Habert and Ouellet-Plamondon, 2016). Overall, fly ash– and slag-based geopolymers exhibit lower AP than OPC, whereas metakaolin-based geopolymers can present higher values depending on the energy source. Reducing dependence on sodium silicate, adopting renewable energy, and valorising industrial waste are key to minimising the impact of acidification.

6.4 EUTROPHICATION POTENTIAL

Environmental impact assessments of construction materials increasingly go beyond greenhouse gas emissions and energy demand to include Eutrophication Potential (EP). Eutrophication refers to the excessive enrichment of water bodies with nutrients, primarily nitrogen and phosphorus, leading to harmful algal blooms, oxygen depletion, biodiversity loss, and ecosystem imbalance. While concrete and cement production are not the largest contributors to global eutrophication, their lifecycle stages (i.e., raw material extraction, fuel combustion, chemical production, and transportation) release pollutants such as nitrates (NO_3^-), nitrogen oxides (NO_x), ammonia (NH_3), and phosphates (PO_4^{3-}). These pollutants commonly originate from chemical industries, combustion processes, the use of fertilizers in agriculture, wastewater discharge, leaching, and mining activities. EP measures the contribution of nutrient emissions to soil and water systems, expressed in kg PO_4^{3-} eq per functional unit. OPC has EP values ranging from 0.003 to 0.006 kg PO_4^{3-} eq per kg of cement, attributed to high-temperature clinker production (releasing NO_x) and to limestone quarrying (leading to PO_4^{3-} runoff). In contrast, geopolymers have EP values ranging from 0.0017 to 0.0051 kg PO_4^{3-} eq per kg binder (Habert and Ouellet-Plamondon, 2016), attributed to the production of alkaline activators, transportation, grinding, kaolin calcination energy, and upstream emissions from clay mining.

EP is a critical but often underexplored environmental category in the sustainability assessment of geopolymers. Compared to OPC, geopolymers – especially those based on fly ash and slag – tend to show lower eutrophication impacts. However, alkaline activator production emerges as a key hotspot, often responsible for the majority of EP in geopolymer systems. Strategies such as waste-derived activators, renewable energy integration, and optimized mix designs offer pathways to reduce eutrophication potential. With these innovations, geopolymers can further strengthen their position as a sustainable alternative to Portland cement, not only in terms of carbon footprint but also across a wider set of environmental indicators.

6.5 ABIOTIC DEPLETION POTENTIAL

Abiotic Depletion Potential (ADP) accounts for the consumption of non-renewable natural resources, including minerals and fossil fuels. ADP is expressed in relation to the scarcity of a given resource and its extraction rate. The crucial contributors to the assessment of ADP comprise mineral resources (e.g., aluminium, iron, copper, kaolin, limestone, gypsum) and fossil fuels (coal, oil, and natural gas). ADP is usually quantified in terms of kilogram antimony equivalents (kg Sb-eq/kg), since antimony is a relatively scarce element used as a reference. For construction materials, ADP becomes critical because large-scale infrastructure development requires significant mineral inputs. Assessing ADP is essential for geopolymers because their raw materials and activators are derived from both industrial by-products and mined natural resources.

FA, GGBFS, and RHA precursors are often considered waste streams from other industries. Their environmental burdens are minimal in terms of resource depletion since they are secondary materials, but allocation methods in LCA may attribute partial impacts depending on the system boundaries. In contrast, metakaolin precursors require calcination of kaolin clay, which entails mining of primary clay deposits. This contributes to mineral resource depletion and significant energy demand. Thus geopolymers relying on industrial by-products generally demonstrate lower ADP (1.05×10^{-5} kg Sb-eq/t) compared to OPC (2.5×10^{-5} kg Sb-eq/t), which relies on virgin materials (Habert and Ouellet-Plamondon, 2016). While geopolymers reduce greenhouse gas emissions, their reliance on alkaline activators may contribute substantially to abiotic depletion. Sodium silicate is the dominant contributor to ADP in geopolymers, often accounting for more than 50–70% of total abiotic depletion, thus highlighting the need for innovative activator alternatives and circular economy strategies.

6.6 COST/BENEFIT ANALYSIS

Geopolymers have gained significant attention as viable alternatives to OPC. Despite their technical, environmental, and durability advantages, their economic viability remains a crucial factor influencing large-scale adoption. The Cost/Benefit Analysis (CBA) of geopolymers involves both direct financial costs and indirect societal benefits (Matsimbe et al., 2022). The direct costs constitute raw material acquisition, transportation, processing, alkali activators, and production energy (McLellan et al., 2011). The indirect costs comprise handling hazardous activators, a limited global large-scale supply chain, and higher initial research and development costs. In contrast, the direct benefits comprise reduced clinker consumption, lower raw material cost, and comparable mechanical performance. The indirect benefits include a reduction in greenhouse gas emissions, waste diversion from landfills, compliance with sustainability regulations, and long-term savings from durability and lower maintenance needs. Increasingly stringent environmental regulations and carbon taxation enhance the economic attractiveness of geopolymers. Furthermore, geopolymers exhibit superior resistance to acid attack, sulphate attack, chloride penetration, and high temperatures compared to OPC-based systems (Matsimbe et al., 2024).

The industrial by-product precursors (e.g., FA, GGBFS, and MT) are often inexpensive or even free in regions with coal power or steel industries (Matsimbe et al., 2024). However, in areas lacking these industrial by-products, procurement can increase costs significantly. The alkaline activators (e.g., NaOH, Na_2SiO_3) are the most expensive components, often accounting for 40–60% of total geopolymer production costs (Matsimbe et al., 2022). Strategies such as partial replacement with locally available natural activators (e.g., rice husk–derived silica) can reduce costs. Geopolymerisation typically occurs at ambient or slightly elevated temperatures, requiring less energy and producing reduced CO_2 emissions compared to ordinary Portland cement clinker (Provis, 2018). However, small-scale production lacks the economies of scale of cement manufacturing, resulting in higher per-unit costs. If waste precursors are sourced locally, transportation costs are minimal.

Furthermore, in regions where materials are imported, logistics add significantly to overall expenses. The development of local geopolymer industries can foster job creation and innovation, especially in areas with abundant agricultural or mining residues. To further advance the cost/benefit profile of geopolymers, strategies should be implemented such as developing alternative low-cost activators; enhancing supply chain integration to reduce logistical costs; scaling up industrial symbiosis with coal, steel, and agricultural industries; instituting government incentives, subsidies, or carbon credits to offset initial higher costs; and establishing design standards and codes to increase trust and reduce premiums.

6.7 END-OF-LIFE OR RECYCLABILITY OF GEOPOLYMERS

The global transition toward sustainable materials requires not only assessing the performance and environmental footprint of materials during production and service life but also understanding their fate at the end of life. Traditional

Portland cement–based concrete presents challenges in recyclability due to its high embodied carbon, contamination risks, and limited reuse options. Geopolymers emerge as promising alternatives due to their durability, reduced CO_2 emissions, and potential for recyclability (Naghizadeh et al., 2025). This section examines the end-of-life pathways of geopolymers, including mechanical, thermal, and chemical recycling, and their potential for integrating the circular economy. LCA emphasises that the environmental impact of construction materials extends beyond raw material extraction and in-service performance. End-of-life (EOL) stages, such as demolition, reuse, recycling, or disposal, can significantly alter the sustainability credentials of a material. For geopolymers, which often utilise industrial by-products such as fly ash or slag (Duxson et al., 2007; Provis and van Deventer, 2009), EOL strategies offer a second opportunity to valorise waste, closing material loops and reducing reliance on landfills (Ntuli et al., 2025).

Like traditional Portland concrete, geopolymer concrete can be crushed into smaller fractions and reused as recycled aggregates in new construction. Crushed geopolymer concrete exhibits high angularity and rough texture, which may enhance the bond with fresh binders. However, the residual alkalinity and high strength of some geopolymer matrices can make crushing more energy-intensive compared to Portland cement concrete. Heating geopolymer waste at elevated temperatures can break down the polymeric network, regenerating reactive aluminosilicate phases (Naghizadeh et al., 2025). This process creates secondary precursors that can be reactivated with alkaline solutions, essentially closing the loop. While thermally recycling geopolymers mimics the production of metakaolin or calcined clays, it demands significant energy input. Its sustainability depends on coupling with renewable energy sources or waste heat recovery from industrial systems.

Chemical recycling targets the dissolution of hardened geopolymer matrices in alkaline solutions to recover reactived silicate and aluminate species. This approach aligns with circular economy strategies but requires further optimisation for scalability and safety. The key aspects to be observed are that:

- the sodium hydroxide solutions can partially dissolve geopolymer gels;
- the dissolved species can be used to form fresh geopolymer pastes;
- closed-loop recycling is theoretically possible, though efficiency is reduced after multiple cycles due to impurities and incomplete dissolution.

Despite its recycling potential, the recycled geopolymer waste must meet the environmental safety standards (Matsimbe et al., 2024). Of course, heavy metals immobilised within the geopolymer matrix remain stable during the recycling process. Nevertheless, leaching tests are critical to avoid groundwater contamination. Compared to Portland cement concrete, geopolymer recycling often requires less processing energy, particularly in mechanical recycling. Additionally, reusing geopolymer aggregates reduces demand for virgin aggregates, minimising land disturbance and CO_2 emissions associated with quarrying. Incorporating the principles of design for construction into geopolymer systems could significantly improve recyclability. For example:

- Modular geopolymer panels that can be disassembled and reused
- Blends optimised for selective dissolution at end of life
- Hybrid composites designed for closed-loop recycling

Geopolymer recycling aligns with broader circular economy frameworks by:

- valorising industrial by-products multiple times;
- reducing dependency on virgin materials;
- supporting the development of urban mining strategies where demolished structures become sources of secondary raw materials.

Despite their recyclability potential, several barriers limit the widespread recycling of geopolymers:

- *Heterogeneity of waste streams:* Different formulations lead to varied recyclability profiles.
- *Lack of standardised recycling protocols:* No universal standards exist for geopolymer EOL handling.
- *Economic viability:* Recycling must compete with landfill disposal and virgin raw materials in cost-effectiveness.
- *Scaling laboratory methods:* Thermal and chemical recycling processes remain largely experimental.

PROBLEMS

1. Explain the four stages of an LCA according to ISO 14040 (ISO 14040, 2006).
2. If producing 1 ton of OPC emits approximately 306 kg CO_2/m^3, and 1 ton of fly ash geopolymer emits 169 kg CO_2/m^3, calculate the percentage reduction in CO_2 emissions.
3. Given the following energy data for 1 m^3 of concrete:
   - *OPC concrete:* 6 MJ/kg of cement
   - *Geopolymer concrete:* 5 MJ/kg of cement

   Compute the energy savings percentage.
4. Discuss why sodium silicate contributes significantly to the environmental impact of geopolymer concrete.
5. Design a cradle-to-grave LCA framework for geopolymer concrete used in bridge construction. Define system boundaries, data sources, and key impact categories.
6. Compare the environmental performance of two geopolymer systems – one using fly ash transported 500 km and another using local slag transported 50 km.
7. Describe how you can develop an uncertainty analysis model to evaluate how variations in precursor chemical composition affect GWP outcomes in geopolymer LCAs.

REFERENCES

Duxson, P., Fernández-Jiménez, A., Provis, J. L., Lukey, G. C., Palomo, A., and van Deventer, J. S. J. (2007). Geopolymer technology: The current state of the art. *Journal of Materials Science,* vol. 42, no. 9, 2917–2933.

Habert, G., and Ouellet-Plamondon, C. (2016). Recent update on the environmental impact of geopolymers. *RILEM Technical Letters,* vol. 1, 17–23.

Hauschild, M. Z., Rosenbaum, R. K., and Olsen, S. I. (2018). *Life Cycle Assessment: Theory and Practice*. Springer Cham.

ISO 14040. (2006). *Environmental Management – Life Cycle Assessment – Principles and Framework.* International Organization for Standardization (ISO).

ISO 14044. (2006). *Environmental Management – Life Cycle Assessment – Requirements and Guidelines*. International Organization for Standardization (ISO).

Matsimbe, J., Dinka, M., Olukanni, D., and Musonda, I. (2022). Geopolymer: A systematic review of methodologies. *Materials*, vol. 15, no. 19.

Matsimbe, J., Dinka, M., Olukanni, D., and Musonda, I. (2024). Durability properties of ambient-cured fly ash-phosphogypsum blended geopolymer mortar in terms of water absorption, porosity, and sulfate resistance. *Discover Sustainability*, vol. 5, no. 330.

McLellan, B. C., Williams, R. P., Lay, J., van Riessen, A., and Corder, G.D. (2011). Costs and carbon emissions for geopolymer pastes in comparison to ordinary Portland cement. *Journal of Cleaner Production*, vol. 19, no. 9–10, 1080–1090.

Naghizadeh, A., Welman-Purchase, M., Lagrange, L., Ekolu, S. O., and Tchadjie, L. N. (2025). Re-activation of recycled fly ash-based geopolymer binder to enhance mechanical properties. *Case Studies in Construction Materials*, vol. 23, e04967.

Ntuli, F., Mashifana, T., Sithole, T., and Falayi, T. (2025). *Municipal Solid, Agricultural, and Mining Waste in Sub-Saharan Africa: Sustainable and Circular Valorization Opportunities* (1st ed.). CRC Press, Boca Raton, FL, USA.

Provis, J. (2018). Geopolymers and other alkali-activated materials: why, how, and what? *Materials and Structures*, vol 51, no. 4, 1–11.

Provis, J., and van Deventer, J. (2009). *Geopolymers: Structures, Processing, Properties and Industrial Applications*. Woodhead Publishing Ltd, Cambridge, UK.

Turner, L. K., and Collins, F. G. (2013). Carbon dioxide equivalent (CO_2-e) emissions: A comparison between geopolymer and OPC cement concrete. *Construction and Building Materials*, vol. 43, 125–130.

Index

For Product Safety Concerns and Information please contact our EU representative GPSR@taylorandfrancis.com
Taylor & Francis Verlag GmbH, Kaufingerstraße 24, 80331 München, Germany

www.ingramcontent.com/pod-product-compliance
Ingram Content Group UK Ltd.
Pitfield, Milton Keynes, MK11 3LW, UK
UKHW021502240926
475702UK00035B/130
9781041308997